W9-DDE-256

Theory of Curve and Surface Evolution:
Corners, Shocks, Singularities and Entropy Conditions
(*Chaps. 2,4: Ref. [225, 222]*)

Tracking Interface Motion with Schemes
from Hyperbolic Conservation Laws
(*Chap. 5: Ref. [226]*)

Level Set Perspective
$\phi_t + F|\nabla\phi| = 0$
Initial Value Problem
(*Chap. 6: Ref. [187]*)

Stationary Perspective
$|\nabla T|F = 1$
Boundary Value Problem
(*Chap. 6: Ref. [99]*)

adaptivity

adaptivity

NARROW BAND
LEVEL SET METHODS

(*Chap. 7: Ref. [2]*)

FAST MARCHING
METHODS

(*Chap. 8: Ref. [233]*)

ADDITIONAL FORMULATIONS

Unstructured Mesh
Level Set Methods
(*Chap. 9: Ref. [24]*)

Coupling to Physics:
Extension Velocities
(*Chap. 11: Ref. [6]*)

Unstructured Mesh
Fast Marching Methods
(*Chap. 10: Ref. [137]*)

APPLICATIONS

Geometry
(Chap. 14)

Grid Generation
(Chap. 15)

Seismic Analysis
(Chap. 20)

Computational Geometry
(Chap. 19)

Computer Vision
(Chaps, 16, 17)

Optimality and Control
(Chap. 20)

Fluid Mechanics
(Chap. 18)

Combustion
(Chap. 18)

Materials Sciences
(Chap. 18)

Semiconductor Manufacturing
(Chap. 21)

This book is an introduction to level set methods and Fast Marching Methods, which are powerful numerical techniques for analyzing and computing interface motion in a host of settings. They rely on a fundamental shift in how one views moving boundaries, rethinking the natural geometric Lagrangian perspective and replacing it with an Eulerian partial differential equation perspectives. The resulting numerical techniques are used to track three-dimensional fronts that can develop sharp corners and change topology as they evolve.

The book begins with an overview of the two techniques, and then provides an introduction to the dynamics of moving curves and surfaces. Next, efficient computational techniques for approximating viscosity solutions to partial differential equations are developed, using numerical technology from hyperbolic conservation laws. This builds a framework for optimal implementations of adaptive techniques for Narrow Band Level Set Methods and Fast Marching Methods. The entire methodology is then redeveloped on triangulated meshes, followed by a series of extensions of the basic ideas. A large collection of applications is given, including examples from physics, chemistry, fluid mechanics, combustion, image processing, materials science, fabrication of microelectronic components, computer vision, control theory, computational geometry, and computer-aided-design and manufacturing.

This book will be a useful resource for mathematicians, applied scientists, practicing engineers, computer graphics artists, and anyone interested in the evolution of boundaries and interfaces.

CAMBRIDGE MONOGRAPHS ON
APPLIED AND COMPUTATIONAL
MATHEMATICS

Series Editors
P. G. CIARLET, A. ISERLES, R. V. KOHN, M. H. WRIGHT

3 # Level Set Methods
and
Fast Marching Methods:

Evolving interfaces in computational geometry,
fluid mechanics, computer vision, and
materials science

The Cambridge Monographs on Applied and Computational Mathematics reflects the crucial role of mathematical and computational techniques in contemporary science. The series publishes expositions on all aspects of applicable and numerical mathematics, with an emphasis on new developments in this fast-moving area of research.

State-of-the-art methods and algorithms as well as modern mathematical descriptions of physical and mechanical ideas are presented in a manner suited to graduate research students and professionals alike. Sound pedagogical presentation is a prerequisite. It is intended that books in the series will serve to inform a new generation of researchers.

Also in this series:

Level Set Methods
and
Fast Marching Methods

Evolving interfaces in computational geometry, fluid mechanics, computer vision, and materials science

J. A. Sethian

University of California, Berkeley

CAMBRIDGE
UNIVERSITY PRESS

PUBLISHED BY THE PRESS SYNDICATE OF THE UNIVERSITY OF CAMBRIDGE
The Pitt Building, Trumpington Street, Cambridge, United Kingdom

CAMBRIDGE UNIVERSITY PRESS
The Edinburgh Building, Cambridge CB2 2RU, UK
40 West 20th Street, New York, NY 10011-4211, USA
10 Stamford Road, Oakleigh, VIC 3166, Australia
Ruiz de Alarcón 13, 28014 Madrid, Spain
Dock House, The Waterfront, Cape Town 8001, South Africa

http://www.cambridge.org

© Cambridge University Press 1996, 1999

This book is in copyright. Subject to statutory exception and
to the provisions of relevant collective licensing agreements,
no reproduction of any part may take place without
the written permission of Cambridge University Press.

First published 1996 *as Leve Set Methods*
Second edition published 1999
Reprinted 2000, 2001

Printed in the United States of America

Typeset in Computer modern 10/13 in LATEX

A catalog record for this book is available from the British Library

Library of Congress Cataloging in Publication Data is available

ISBN 0 521 64557 3 paperback

Contents

vii

Contents

Preface to the Second Edition

The beginning of this work on Level Set Methods and Fast Marching Methods can be found in the author's dissertation on the theory and numerics of propagating interfaces, under the direction of Alexandre Chorin at the University of California at Berkeley. That work continued through a National Science Foundation (NSF) Postdoctoral Fellowship at the Lawrence Berkeley National Laboratory (LBNL) and the Courant Institute of Mathematical Sciences. As I look back on those years, I am extraordinarily grateful for the opportunity that such support gave me to develop these ideas, free from other burdens and responsibilities.

That work on interface methods and its subsequent development at Berkeley have been supported in part by the Applied Mathematical Sciences section of the Department of Energy through the Mathematics Department at LBNL, by NSF awards through the University of California at Berkeley Mathematics Department, and most recently through the Office of Naval Research. Again, I am grateful for all of this support.

I have had the good fortune to work with many collaborators in the development of these ideas. The time-dependent level set formulation of these ideas on interface motion was co-authored with S. J. Osher, whose trips to Berkeley made for a thoroughly enjoyable collaboration. The Narrow Band Level Set Method was developed jointly with D. Adalsteinsson, as was all the work on etching and deposition in semiconductor manufacturing. The work on medical imaging and shape segmentation was joint with R. Malladi, whose help on devising the Fast Marching Method was invaluable. Finally, the triangulated unstructured versions of the Level Set Method are joint with T. Barth of National Aeronautics and Space Administration (NASA) Ames Research Center.

An early application of these techniques, due to D. Chopp, concerns minimal surfaces and includes the genesis of ideas about narrow banding and complex boundary conditions. The work on Level Set Methods for crystal growth and dendritic solidification is joint with J. Strain and capitalizes on his boundary integral formulation of the equations of motion. The realization that level set techniques can be applied to shape recovery is due to R. Malladi; many others have capitalized on his ideas. The application of these techniques to image processing began with the work of L. Alvarez, J. M. Morel, and P. L. Lions and the work of S. Osher and L. Rudin; the work on image processing presented here relies heavily on those contributions. The application of the techniques to problems in combustion and fluid interfaces discussed here is joint work with C. Rhee, J. Zhu and L. Talbot, and the work on adaptive mesh refinement relies on the work of B. Milne.

I am fortunate to have on-going collaborations with very talented colleagues, including M. Popovici of 3DGeo Corporation with whom the work on migration and seismic imaging was developed; R. Kimmel, with whom the work on geodesics and the development of a triangulated Fast Marching Method was developed, as well as robotic navigation and optimal path planning; O. Hald, whose analysis of non-convex Hamiltonians was invaluable; as well as L. Borucki and J. Rey in a wide collection of semiconductor issues. At the same time, I am equally fortunate to have been recently benefited from new collaborations with J. Li, A. Sarti, A. Vladimirsky, J. Wei, A. Wiegmann, and J. Wilkening.

Many other people have contributed to the current state-of-the-art of level set methods, including T. Aslam, M. Barlaud, J. Bence, M. Brewer, J. Bzdil, R. Caflisch, V. Caselles, T. Chan, S. Chen, Y. Chen, R. Deriche, V. Dhir, E. Fatemi, O. Faugeras, R. Fedkiw, E. Harabetian, E. Holm, T. Hou, R. Keriven, H.P. Langtangen, D. Lesselier, A. Litman, B. Merriman, E. Pasch, V. Prasad, S. Ruuth, F. Santosa, P. Smereka, N. Sochen, G. Son, S. Stewart, M. Sussman, H. Zhang, H. Zhao, and L.L. Zheng. Their work has advanced both the theory and practice of level set methods. I would like to thank W. Coughran for suggesting the application of level set methods to semiconductor simulations, A. Neureuther for many helpful discussions on etching and deposition, B. Knight for encouragement in the application of level set methods to fluid interface problems, C. Ritchie and G. Chiang for their insightful suggestions about shape recovery in medical imaging, T. Baker for helpful conversations about grid generation, L. Gray for suggesting the application of level set methods to material sintering, and C. Evans for his valuable comments on the initial manuscript. I also wish to thank the students in Math 273 during the fall of 1995 for their insightful comments, critical reviewing, and careful suggestions.

I am also indebted to the many readers of the first edition who have responded with detailed suggestions, and I have been guided by their comments in writing this new edition. At the risk of repetition, I would like to again thank D. Adalsteinsson, D. Chopp, R. Kimmel, R. Malladi, B. Milne, A. Vladimisky, J. Wilkening and J. Zhu. I am extraordinarily fortunate to have had them as colleagues at the Lawrence Berkeley Laboratory.

Finally, I would like to thank Alan Harvey of Cambridge University Press for his thoughtful suggestions and enthusiasm for this project. His calm hand and unfailing humor have made this a pleasure, and his guidance, wise counsel, and wisdom were invaluable. This second and expanded edition is largely due to his optimism, encouragement, and patience.

Berkeley, California, 1999

Introduction

Propagating interfaces occur in a wide variety of settings, and include ocean waves, burning flames, and material boundaries. Less obvious boundaries are equally important and include shapes against backgrounds, handwritten characters, and iso-intensity contours in images. Furthermore, there are applications not commonly thought of as moving interface problems, including optimal path planning and construction of shortest geodesic paths on surfaces, which can be recast as front propagation problems with significant advantages.

The goal of this book is both to unify these ideas and to design a general framework for modeling the evolution of boundaries. The aim is to provide computational techniques for tracking moving interfaces and to give some hint of the flavor and breadth of applications. The work includes examples from physics, chemistry, fluid mechanics, combustion, image processing, materials sciences, fabrication of microelectronic components, computer vision, control theory, seismology, computer-aided-design, and a collection of other areas. The intended audience includes mathematicians, applied scientists, practicing engineers, computer graphics artists, and anyone interested in the evolution of boundaries and interfaces.

Our perspective comes from a large and rapidly growing body of work which relies on a partial differential equations approach for understanding, analyzing, and computing interface motion. At the core lay two computational techniques: "Fast Marching Methods" and "Level Set Methods". Both exploit a fundamental shift in how one views moving boundaries. They rethink the Lagrangian geometric perspective and replace it with an Eulerian, partial differential equation. Fast Marching Methods result from a boundary value problem for the evolving interface, while Level Set Methods result from an associated initial value problem.

In both cases, several advantages stem from this view of propagating interfaces:

- First, from a theoretical/mathematical point of view, some complexities of front motion are illuminated, in particular, the role of singularities, weak solutions, shock formation, entropy conditions, and topological change in the evolving interface.

- Second, from a numerical perspective, natural and accurate ways of computing delicate quantities emerge, including the ability to build high order advection schemes, compute local curvature in two and three dimensions, track sharp corners and cusps, and handle subtle topological changes of merger and breakage.

- Third, from an implementation point of view, since the approaches are based on underlying partial differential equations, robust schemes result from numerical parameters set at the beginning of the computation. The error is thus controlled by

  (i) the order of the numerical method,

  (ii) the grid spacing Δh,

  (iii) in the case of Level Set Methods, the time step Δt; no such requirement exists for Fast Marching Methods.

- Fourth, computational adaptivity is the key to these techniques. In the case of Level Set Methods, the most efficient and preferred approach is the "Narrow Band Level Set Method", which focuses computational labor around the evolving boundary. In the case of Fast Marching Methods, use of standard sorting techniques yields extraordinarily fast and optimally efficient algorithms. In both cases, a clear path to parallelism is available.

This book surveys what is intended to be an illustrative subset of past and current applications of these techniques. We do not assume that the reader is familiar with all of the details required to develop these schemes; the aim is to include the necessary theory and details to provide implementation guidelines.

The first edition of this book was entitled <u>Level Set Methods</u>. The augmented title <u>Level Set Methods and Fast Marching Methods</u> of this new edition embraces the large landscape shared by these two techniques in framing, illuminating, and solving problems with evolving boundaries.

Outline

This book is divided into four parts. Part I focuses on the formulation of the boundary value and initial value partial differential equations which comprise our two views of interface motion. Part II introduces the theory and numerics underlying Fast Marching Methods and Level Set Methods. Part III introduces the adaptive issues required to construct efficient schemes and variations on the fundamental techniques. Finally, Part IV surveys some application areas.

In **Part I**, Chapter 1 begins with the underlying boundary value and initial value partial differential equations perspective on moving interfaces, and discusses the theoretical and computational advantages of these approaches. It ends with a preview of the rest of the book and provides an outline of the interconnection of the techniques, the relevant theory, numerics, and application strategies. This "look ahead" is meant to provide a structure for the remainder of the book, directing the interested reader to various components of the methodologies.

Part II begins in Chapter 2 with a general statement of the problem of a moving interface and discusses the mathematical theory of curve/surface motion, including the growth/decay of total variation, singularity development, entropy conditions, weak solutions, and shocks in the dynamics of moving fronts. This material has been developed in a collection of papers that are referred to in the text. The viscosity theory of Hamilton-Jacobi equations, which buttresses both computational techniques, is briefly surveyed in Chapter 3.

Chapters 4, 5, and 6 present numerical results which lead up to the Fast Marching and Level Set techniques. Chapter 4 begins with an overview of traditional methods for tracking interfaces, including string methods and cell methods, and makes a first attempt at solving a partial differential equation for front propagation The failure of this first attempt stems from the relationship between front propagation and hyperbolic conservation laws and is the subject of Chapter 5. Chapter 6 then provides a detailed description of straightforward (though inefficient) algorithms for solving the initial value and boundary value problems.

Part III provides complete details on state-of-the-art Fast Marching and Level Set algorithms. It begins in Chapter 7 with a discussion of computational adaptivity. After surveying work on parallel and adaptive mesh approaches, the chapter focuses on the Narrow Band Level Set Method. This is the most efficient and accurate way to implement level set methods. Next, in Chapter 8, Fast Marching Methods are introduced, which are the optimal way to solve Hamilton-Jacobi equations which arise from certain interface motion problems. The techniques require a detailed discussion of causality in upwind schemes and optimal heap sort algorithms. Higher accuracy versions of both Narrow Band and Fast Marching Methods are supplied.

Next, in Chapters 9 and 10, the entire framework is moved to a triangulated unstructured mesh setting. Schemes for the Level Set Method are given, including monotone schemes, positive schemes, Petrov-Galerkin schemes, as well as explicit and implicit schemes with discontinuity capturing. In the case of Fast Marching Methods, upwind causality schemes for both acute and non-acute triangulations are introduced. These two sets of schemes provide versatile techniques for interface propagation problems on manifolds and in irregular domains.

Chapter 11 explains how to build general level set methods in many physical problems. It examines how to build appropriate and natural methods for moving the neighboring level sets, which is required in order to implement level set techniques. Detailed techniques for generating smooth level set flows which avoids all re-initialization are given, as are techniques for obtaining sub-grid accuracy. In Chapter 12, the numerical accuracy and robustness tests are measured, including scheme convergence rates, tests of triangulated techniques, examination of mass conservation and accuracy. Finally, in Chapter 13, the underlying philosophy of Narrow Band and Fast Marching Methods applications is discussed.

Part IV focuses on applications of both the Narrow Band Level Set Method and Fast Marching Methods to a collection of problems. Here, the intent is to show the breadth of current applications and to serve as a guidepost for further research. Chapter 14 begins with some pure geometry problems, including curve/surface shrinkage, the existence of self-similar surfaces, flows under more complex metrics, sintering and second derivative of curvature flows, triple points, multiple interfaces, and constraint-based flows. Chapter 15 extends this work and shows how these techniques can be used in grid generation, giving many examples of body-fitted logical rectangular grids around complex bodies in two and three dimensions. Chapter 16 moves to image processing and views images as collections of iso-intensity contours; by constructing a suitable speed law, these contours can be allowed to propagate in a way that both removes noise and enhances desired regions.

Chapter 17 focuses on aspects of computer vision. It begins with the problem of shape-from-shading and then shows how to transform image segmentation problems into moving interface versions of active contours; when driven by gradients in the image field, these contours extract desired shapes from images. Applications are drawn from a wide collection of medical data, including three-dimensional scans of cortical and cardiac structures. Once images are segmented, the next step is recognition, which is discussed in the context of automatic identification of meteorological data and optical character recognition.

Chapter 18 provides examples of interface problems in which the physics on each side of the interface both drives the front and is affected by the front location and properties. Several areas are discussed, including combustion and flame propagation, crystal growth and dendritic solidification, fluid interface transport and two-phase flow, and electromigration. In all four, the front is driven both by local effects and by underlying transport terms. General guidelines for arbitrary fluid/material interface problems are given. Additional applications include boiling, groundwater transport, and liquid bridges.

Chapter 19 focuses on various aspects of computational geometry and computer-aided-design. It begins with efficient algorithms for shape-offsetting. Next, techniques for constructing minimal surfaces are given which rely on constrained fronts evolving under mean curvature until final minimal steady states are achieved. The chapter ends with problems of shape smoothing, of importance in removing noise from range images as well as machine part manufacturing. This is performed using variants of the image smoothing schemes presented earlier.

Chapter 20 applies Fast Marching methods to a variety of problems in computing first arrivals, optimization, and control. It begins with problems in path planning and navigation under constraints. It then gives an optimal algorithm for constructing shortest path geodesics on complex triangulated manifolds, including a technique for ruling surfaces. It then discusses first arrival times of seismic waves in geophysical migration modeling and problems applying level set methods to air-traffic control. The chapter ends with some new algorithms for computing visibility in complex scenes.

Chapter 21 presents the most sophisticated application of Fast Marching/Level Set methodologies to date, namely the simulation of etching, deposition, and photolithography development in the microfabrication of semiconductor components. Here, photolithography development in planar and non-planar domains, etching and deposition with non-convex ion-milling sputter effects, re-emission and re-deposition mechanisms with small sticking coefficients, passive sidewall activation, surface diffusion and re-flow, as well as full three-dimensional effects are discussed. This requires the use of efficient visibility schemes, schemes for sintering, fast solution of flux integral equations, and sub-grid adaptivity schemes.

By no means is this an exhaustive review of the work that exists on Fast Marching Methods and Level Set Methods. A body of work has been reluctantly skipped in the effort to keep this book to reasonable length. The interested reader is referred to a wide range of simulations developed using these methodologies; references will be given throughout the text. The goal of this book is to provide windows into these techniques as guides for further interface studies.

The author can be reached at sethian@math.berkeley.edu. A general article on Fast Marching Methods and Level Set Methods may be found in [237], other reviews may be found in [238] and [235]. Finally, a web page devoted to the topic of Fast Marching Methods and Level Set Methods may be found at http://math.berkeley.edu/~sethian/level_set.html.

Part I

Equations of Motion for Moving Interfaces

Part I presents the underlying partial differential equations perspective on moving interfaces. One view leads to a boundary value partial differential equation for the evolving front, the other leads to a time-dependent initial value problem. The goal is to lay out clearly the two views and discuss the theoretical and computational advantage of these approaches.

1

Formulation of Interface Propagation

Outline: *We formulate the boundary value and initial value partial differential equations which describe interface motion. These will eventually lead to the Fast Marching Method and the Narrow Band Level Set Method; for now, however, we focus on the theoretical and computational advantages that come from these perspectives.*

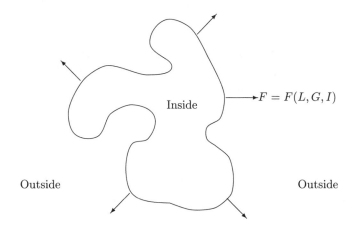

Fig. 1.1. Curve propagating with speed F in normal direction.

Consider a boundary, either a curve in two dimensions or a surface in three dimensions, separating one region from another. Imagine that this curve/surface moves in a direction normal to itself (where the normal direction is oriented with respect to an inside and an outside) with a known speed function F. The goal is to track the motion of this interface as it evolves. We are concerned only with the motion of the interface in

3

its normal direction; throughout, we shall ignore motions of the interface in its tangential directions.

The speed function F, which may depend on many factors, can be written as:

$$F = F(L, G, I), \qquad (1.1)$$

where

- L= Local properties are those determined by local geometric information, such as curvature and normal direction.
- G= Global properties of the front are those that depend on the shape and position of the front. For example, the speed might depend on integrals along the front and/or associated differential equations. As a particular case, if the interface is a source of heat that affects diffusion on either side of the interface, and a jump in the diffusion in turn influences the motion of the interface, then this would be characterized as global property.
- I= Independent properties are those that are independent of the shape of the front, such as an underlying fluid velocity that passively transports the front.

Much of the challenge in interface problems comes from producing an adequate model for the speed function F; this is a separate issue independent of the goal of an accurate scheme for advancing the interface based on the model for F. In this chapter, it is assumed that the speed function F is known. The goal of Part IV is to formulate good models for F for a collection of applications.

Given F and the position of an interface, the objective is to track the evolution of the interface. Our first task is to formulate this evolution problem in an *Eulerian* framework, that is, one in which the underlying coordinate system remains fixed.

1.1 A boundary value formulation

Assume for the moment that $F > 0$, hence the front always moves "outward." One way to characterize the position of this expanding front is to compute the arrival time $T(x, y)$ of the front as it crosses each point (x, y), as shown in Figure 1.2.

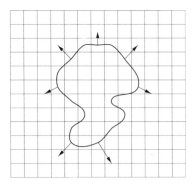

Fig. 1.2. Calculation of crossing time at (x, y) for expanding front $F > 0$.

The equation for this arrival function $T(x, y)$ is easily derived. In one dimension, using the fact that *distance* = *rate* $*$ *time* (see Figure 1.3), we have that

$$1 = F \frac{dT}{dx}.$$

$$F = dx/dT$$

$$x = FT$$

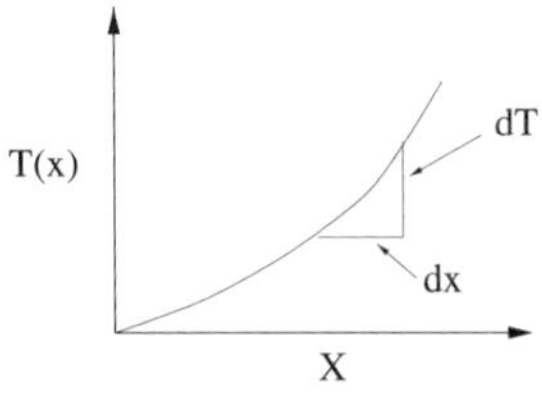

T(x)

dT

dx

X

Fig. 1.3. Setup for boundary value formulation.

In multiple dimensions, ∇T is orthogonal to the level sets of T, and, similar to the one-dimensional case, its magnitude is inversely proportional to the speed. Hence

$$|\nabla T|F = 1, \qquad T = 0 \text{ on } \Gamma, \qquad (1.2)$$

where Γ is the initial location of the interface.

Thus, the front motion is characterized as the solution to a boundary value problem. If the speed F depends only on position, then the equation reduces to what is known as the "Eikonal" equation. As an example, the arrival surface $T(x, y)$ for a circular front expanding with unit speed $F = 1$ is shown in Figure 1.4.

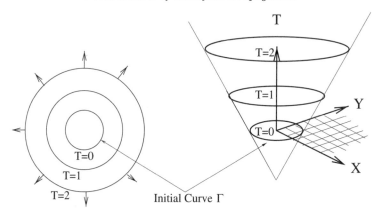

Fig. 1.4. Transformation of front motion into boundary value problem.

1.2 An initial value formulation

Conversely, suppose now that the front moves with a speed F that is neither strictly positive nor negative. Then we must account for the fact that the front can move forward and backward, and hence can pass over a point (x, y) several times. Thus, the crossing time $T(x, y)$ is not a single-valued function. Our way of taking care of this is to embed the initial position of the front as the zero level set of a higher-dimensional function ϕ. We can then link the evolution of this function ϕ to the propagation of the front itself through a time-dependent initial value problem. At any time, the front is given by the zero level set of the time-dependent level set function ϕ (see Figure 1.5).

In order to derive an equation of the motion for this level set function ϕ and match the zero level set of ϕ with the evolving front, we first require that the level set value of a particle on the front with path $x(t)$ must always be zero, and hence

$$\phi(x(t), t) = 0. \tag{1.3}$$

By the chain rule,

$$\phi_t + \nabla\phi(x(t), t) \cdot x'(t) = 0. \tag{1.4}$$

Since F supplies the speed in the outward normal direction, then $x'(t) \cdot n = F$, where $n = \nabla\phi/|\nabla\phi|$. This yields an evolution equation for ϕ, namely

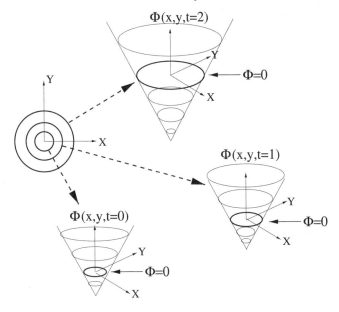

Fig. 1.5. Transformation of front motion into initial value problem.

$$\phi_t + F|\nabla \phi| = 0, \qquad (1.5)$$

$$\text{given} \quad \phi(x, t = 0).$$

This is the level set equation given by Osher and Sethian [187]. For certain forms of the speed function F, one obtains a standard Hamilton–Jacobi equation. Equation 1.5 describes the time evolution of the level set function ϕ in such a way that the zero level set of this evolving function is always identified with the propagating interface; see Figure 1.5.

Thus, we can summarize our two perspectives. Let Γ be a curve in the plane propagating in a direction normal to itself with speed F such that $\Gamma(t)$ gives the position of the front at time t. Then, we wish to solve

Boundary Value Formulation	**Initial Value Formulation**
$|\nabla T|F = 1$	$\phi_t + F|\nabla \phi| = 0$
Front= $\Gamma(t) = \{(x,y)|T(x,y) = t\}$	Front=$\Gamma(t) = \{(x,y)|\phi(x,y,t) = 0\}$
Requires $F > 0$	Applies for arbitrary F

$$(1.6)$$

1.3 Advantages of these perspectives

There are certain advantages associated with these two perspectives on propagating interfaces.

- Both are unchanged in higher dimensions, that is, for hypersurfaces propagating in three dimensions and higher.

- Topological changes in the evolving front Γ are handled naturally. The position of the front at time t is given either by the zero level set $\phi(x, y, t) = 0$ of the evolving function ϕ or by the level set $T(x, y) = t$ of the boundary value solution. This set need not be a single curve, and it can break and merge as t advances. In both cases, the key fact is that the boundary value solution $T(x, y)$ and the level set function ϕ remain single-valued (see Figure 1.6).

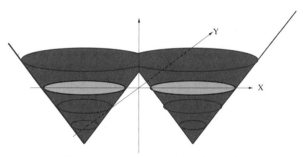

The level set surface ϕ (dark gray):
Two separate initial fronts (in light gray).

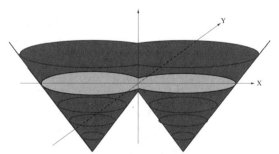

Later in time: the interface topology has changed,
yielding a single curve as the zero level set.

Fig. 1.6. Topological change.

- Both rely on viscosity solutions of the associated partial differential equations in order to guarantee that the unique, entropy-satisfying weak solution is be obtained.

- Both are accurately approximated by computational schemes which exploit techniques borrowed from the numerical solutions of hyperbolic conservation laws. For example, schemes may be developed by using a discrete grid in x-y domain and substituting finite difference approximations[1] for the spatial and temporal derivatives. As illustration, using a uniform mesh of spacing h, with grid nodes (i, j), and employing the standard notation that ϕ_{ij}^n is the approximation to the solution $\phi(ih, jh, n\Delta t)$, where Δt is the time step, one might write

$$\frac{\phi_{ij}^{n+1} - \phi_{ij}^n}{\Delta t} + F|\nabla_{ij}\phi_{ij}^n| = 0. \tag{1.7}$$

  Here, a forward difference scheme in time has been used, and $|\nabla_{ij}\phi_{ij}^n|$ represents some appropriate finite difference operator for the spatial derivative. Thus, an explicit finite difference approach is possible. The construction of correct entropy-satisfying approximations to the difference operator is the subject of Part II; for now, the important fact is that one has an explicit error control on the basis of the initial spatial discretization and the order of the numerical scheme.

- Intrinsic geometric properties of the front are easily determined in both formulations. For example, at any point of the front, the normal vector is given by

  normal to the surface see Edwards + Penney (Calculus)

$$\vec{n} = \frac{\nabla \phi}{|\nabla \phi|} \quad \text{or} \quad \vec{n} = \frac{\nabla T}{|\nabla T|}, \tag{1.8}$$

  and the curvature of the front at any point is easily obtained from the divergence of the unit normal vector to the front, i.e.,

$$\kappa = \begin{cases} \nabla \cdot \dfrac{\nabla \phi}{|\nabla \phi|} = \dfrac{\phi_{xx}\phi_y^2 - 2\phi_x\phi_y\phi_{xy} + \phi_{yy}\phi_x^2}{(\phi_x^2 + \phi_y^2)^{3/2}} \\[2ex] \nabla \cdot \dfrac{\nabla T}{|\nabla T|} = \dfrac{T_{xx}T_y^2 - 2T_xT_yT_{xy} + T_{yy}T_x^2}{(T_x^2 + T_y^2)^{3/2}} \end{cases}. \tag{1.9}$$

- Both methods are made efficient through the use of adaptive computational strategies, which lead to Narrow Band Level Set Methods and Fast Marching Methods.

---

[1] Finite difference approximations will be discussed in detail in Chapter 5.

At the same time, there are significant differences between the two approaches.

- The most obvious difference is that the initial value level set formulation allows for both positive and negative speed functions F; the front may move forward and backward as it evolves. The boundary value perspective is restricted to fronts that always move in the same direction, because it requires a single crossing time T at each grid point, and hence a point cannot be revisited. Thus, models involving more complex speed functions F, such as those including curvature, are most naturally framed as initial value level set problems.

- Conversely, positive speed functions F which depend on position and vary widely from point to point are best framed as boundary value problems and approximated through the use of Fast Marching Methods. This is because

  (i) The boundary value formulation requires no time step, and hence its approximation is not subject to CFL conditions, unlike Level Set Methods.

  (ii) Through the use of heap sort algorithms, Fast Marching Methods can be made extremely computationally efficient, far eclipsing Level Set Methods.

$$|\nabla T| F = 1 \qquad \phi_t + F|\nabla\phi| = 0$$

$$T = u$$

1.4 A general framework

$$\phi = u$$

$$F = |\nabla u|$$

We can be slightly informal and describe both formulations with the general partial differential equation

$$\alpha u_t + H(Du, x) = 0. \tag{1.10}$$

Here, Du represents the partials of u in each variable, for example, u_x and u_y. In the case of the Eikonal equation, $\alpha = 0$, and the function H reduces to $H = F|\nabla u| - 1$.

One of the main subtleties that arises in solving this equation is that the solution need not be differentiable, even with arbitrarily smooth boundary data. This non-differentiability is intimately connected to the notion of appropriate weak solutions. Our goal will be to construct numerical techniques which naturally account for this non-differentiability in the construction of accurate and efficient approximation schemes and admit physically correct non-smooth solutions.

1.5 A look ahead/A look back

It is worthwhile to stop and explain how these techniques were developed and what lies ahead. The first step in the development of these ideas started with the analysis of corners and singularities in propagating interfaces. In [222, 225], the role of curvature as a regularizing or smoothing term was investigated, and it was shown that this regularizing role connects to the notion of entropy conditions and shocks in hyperbolic conservation laws in gas dynamics. This is the subject of Chapter 2. A more formal view comes from considering viscosity solutions of Hamilton-Jacobi equations, which is the subject of Chapter 3.

The second step in the development of accurate and efficient numerical techniques for interface evolution comes from the realization that the schemes from computational fluid mechanics, specifically designed for approximating the solution to hyperbolic conservation laws, can be used to solve the equations of front propagation. This was the view developed in [226], and is at the core of modern interface methods:

> "Most algorithms place marker particles along the front and advance the position of the particles in accordance with a set of finite difference approximations to the equations of motion. Such schemes usually go unstable and blow up as the curvature builds around a cusp, since small errors in the position produce large errors in the determination of the curvature. One alternative is to consider the reformulation equations of motion as a conservation law with viscosity and solve these equations with the techniques developed for gas dynamics. These techniques, based on high-order upwind formulations, are particularly attractive, since they are highly stable, accurate and preserve monotonicity. We have made some preliminary tests of such schemes applied to our problem of propagating fronts in crystals and flames, with extremely encouraging results..."

To execute this strategy, we need schemes from hyperbolic conservation laws; this is the subject of Chapters 4 and 5.

The combination of these three subjects then leads to the two numerical schemes given in Chapter 6: the Level Set Method ([187]) for the initial value problem, and an iterative method for the boundary value problem. They are made efficient in Chapters 8 and 9 through adaptivity, leading to Narrow Band Level Set Methods, see [2], and Fast Marching Methods, see [233]. Finally, after a series of extensions of the basic techniques are developed in Part III, many applications are described in Part IV.

The interconnectedness of past work on methods for interface propagation and the set of applications to be discussed are shown in Figure 1.7. There are many other contributors to the evolution of these ideas; the chart is meant to give perspective on how the theory, algorithms, and applications have evolved. We urge the reader to consult the bibliography to get a more complete sense of the literature and the range of work underway.

1.6 A larger perspective

Fast Marching Methods and Level Set Methods offer powerful techniques for tracking moving interfaces. This book (like its previous edition) aims to demonstrate how these techniques are applied across a wide spectrum of applications. However, there are many other ways to compute solutions to these problems besides the techniques offered here. Marker particle techniques, Volume–of–fluid simulations, Fourier techniques, and phase field models all offer valuable approaches. For each application area given, there is a substantial literature which describes other approaches. At the same time, new techniques and algorithms are always under development, and one of the surest ways to render a body of work obsolete is to pronounce that it can't be improved upon. With that in mind, our goal here is to capture the flavor, intuitive feel, and details of Fast Marching and Level Set Methods.

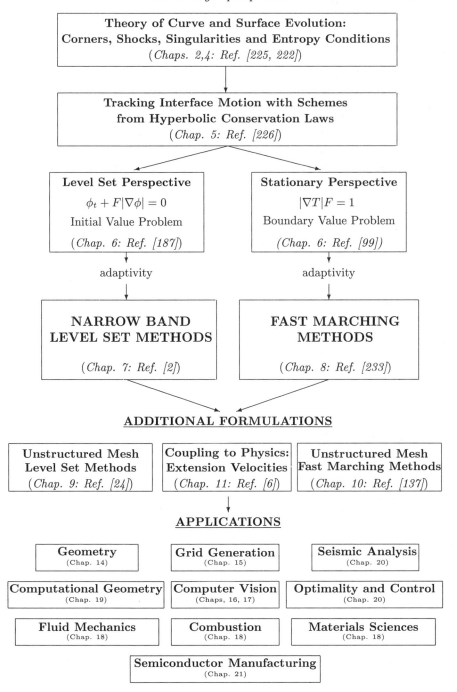

Fig. 1.7. Evolution of theory and algorithms for interface propagations.

Part II

Theory and Algorithms

Part II begins with a theoretical analysis of propagating interfaces, with the goal of analyzing the stability and smoothness of solutions as functions of initial position and speed. We discuss both Lagrangian and Eulerian formulations of the equations of motion. This leads to the link between propagating interfaces and hyperbolic conservation laws, as well as a discussion of singularities and weak solutions. Formally, this connects to the theory of viscosity solutions, which is summarized. We then discuss traditional schemes for tracking interfaces, which are followed by a review of numerical schemes for conservation laws. Finally, basic schemes for our initial and boundary value formulations of interface motion are given.

2

Theory of Curve and Surface Evolution

Outline: *We formulate the equations of motion of a propagating curve, study its stability, and show that corners (singularities in the curvature) can develop as the front evolves. We then show that these corners are analogous to shocks in the solution of hyperbolic conservation laws and that a solution can be naturally constructed beyond the appearance of these corners through the notion of an entropy-satisfying weak solution.*

2.1 Fundamental formulation

Let γ be a simple, smooth, closed initial curve in R^2, and let $\gamma(t)$ be the one-parameter family of curves generated by moving γ along its normal vector field with speed F. Here, F is the given scalar function. Thus $\vec{n} \cdot \vec{x}_t = F$, where $\vec{x}$ is the position vector of the curve, t is time, and $\vec{n}$ is the unit normal to the curve.

A natural approach is to consider a parameterized form of the equations. In this discussion, we further restrict ourselves and imagine that the speed function F depends only on the local curvature κ of the curve, that is, $F = F(\kappa)$.[1] Let the position vector $\vec{x}(s,t)$ parameterize γ at time t, where $0 \leq s \leq S$, and assume periodic boundary conditions $\vec{x}(0,t) = \vec{x}(S,t)$. The curve is parameterized so that the interior is on the left in the direction of increasing s (see Figure 2.1). Let $\vec{n}(s,t)$ be the parameterization of the outward normal and let $\kappa(s,t)$ be the parameterization of the curvature. The equations of motion can then be

---

[1] Curvature is a vector that points in the direction normal to the curve; since we are always taking the speed function in the normal direction, we shall abuse notation and write, for example, $F = -\kappa$. With our sign convention, a counterclockwise parameterized curve has positive curvature.

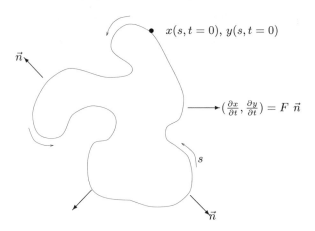

Fig. 2.1. Parameterized view of propagating curve.

written in terms of individual components $\vec{x} = (x, y)$ as

$$x_t = F \left[\frac{y_{ss}x_s - x_{ss}y_s}{(x_s^2 + y_s^2)^{3/2}} \right] \left(\frac{y_s}{(x_s^2 + y_s^2)^{1/2}} \right),$$

$$y_t = -F \left[\frac{y_{ss}x_s - x_{ss}y_s}{(x_s^2 + y_s^2)^{3/2}} \right] \left(\frac{x_s}{(x_s^2 + y_s^2)^{1/2}} \right), \qquad (2.1)$$

where we have used the parameterized expression $\kappa = \frac{y_{ss}x_s - x_{ss}y_s}{(x_s^2 + y_s^2)^{3/2}}$ for the curvature inside the speed function $F(\kappa)$ and the fact that the normal is given by $\vec{n} = (y_s, -x_s)/(x_s^2 + y_s^2)^{1/2}$. This is a "Lagrangian" representation because the range of $(x(s,t), y(s,t))$ describes the moving front.

2.2 Total variation: stability and the growth of oscillations

What happens to oscillations in the initial curve as it moves? We summarize the argument by Sethian [225] showing that the decay of oscillations depends only on the sign of F_κ at $\kappa = 0$. Recall that the metric $g(s,t) = (x_s^2 + y_s^2)^{1/2}$ measures the "stretch" of the parameterization. Define the total oscillation (also known as the total variation) of the front

$$\text{Var}(t) = \int_0^S |\kappa(s,t)| g(s,t) ds. \qquad (2.2)$$

Without the absolute value sign around the curvature, this evaluates to 2π; the absolute value sign allows $\text{Var}(t)$ to measures the amount of

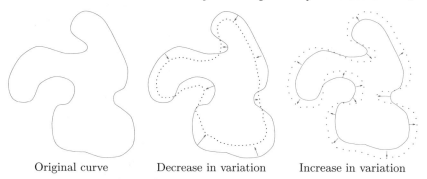

Fig. 2.2. Change in variation Var(t).

"wrinkling". Our goal is to see if this wrinkling increases or decreases as the front evolves; two possible flows are shown in Figure 2.2.

Differentiation of the curvature and the metric with respect to time, together with substitution from Eqn. 2.1 produces[2] the corresponding evolution equations for the metric and curvature, namely,

$$\kappa_t = -g^{-1}(F_s g^{-1})_s - \kappa^2 F, \qquad (2.3)$$

$$g_t = g\kappa F. \qquad (2.4)$$

(Here, g^{-1} is $1/g$, not the inverse.) Suppose we have an initial curve moving with speed $F(\kappa)$ which stays smooth. Evaluation of the time change of the total variation in the solution yields the following [225]:

Proposition: Consider a front moving along its normal vector field with speed $F(\kappa)$, as in Eqn. 2.1. Assume that the evolving curve $\gamma(t)$ is simple, three times differentiable for $0 \leq s \leq S$ and $0 \leq t \leq T$, and non-convex, so that $\kappa(s,0)$ changes sign. Assume that F is twice differentiable. Then, for $0 \leq t \leq T$,

- if $F_\kappa(0) \leq 0$ $(F_\kappa(0) \geq 0)$, then

$$\frac{d\,\mathrm{Var(t)}}{dt} \leq 0 \quad \left(\frac{d\,\mathrm{Var(t)}}{dt} \geq 0 \right), \qquad (2.5)$$

- if $F_\kappa(0) < 0$ $(F_\kappa(0) > 0)$ and $\kappa_s(0) \neq 0$, then

$$\frac{d\,\mathrm{Var(t)}}{dt} < 0 \quad \left(\frac{d\,\mathrm{Var(t)}}{dt} > 0 \right). \qquad (2.6)$$

---

[2] With some work!

Remarks: The proposition states that if $F_\kappa < 0$ wherever $\kappa = 0$, then the total variation decreases as the front moves and the front "smooths out," that is, the energy of the front dissipates. Here, we assume that the front remains twice differentiable in the interval $0 \le t \le T$; in the next section, we discuss what happens if the front ceases to be smooth and develops a corner. In the special case that $\gamma(t)$ is convex for all t, the proposition is trivial, since $\mathrm{Var}(t) = \int_0^S \kappa g ds = 2\pi$.

Outline of proof of proposition: To prove the proposition, the integral is broken up into sections where the curvature changes sign. The time differentiation of the total variation can then be passed to each section of the curve. Using the expressions for the time derivatives of both the metric and the curvature and noting the change in sign of the curvature κ from one section to the next, the decay or growth in the total variation can be evaluated. A full proof may be found in [225].

Two important cases can be easily checked. A speed function $F(\kappa) = 1 - \epsilon\kappa$ for $\epsilon > 0$ has derivative $F_\kappa = -\epsilon$, and hence the total variation decays. Conversely, a speed function of the form $F(\kappa) = 1 + \epsilon\kappa$ yields a positive speed derivative, and hence oscillations grow. We shall see that the sign of the curvature term in this case corresponds to the backward heat equation and hence must be unstable.

2.3 The role of entropy conditions and weak solutions

The above proposition includes the assumption that the front stays smooth. In all but the simplest flows, this smoothness is soon lost. For example, consider the periodic initial cosine curve

$$\gamma(0) = (1 - s, [1 + \cos 2\pi s]/2) \qquad (2.7)$$

propagating with speed $F(\kappa) = 1$. (The parameterization is chosen so that the inside is on the left as we move in the direction of increasing s.) The exact solution to this problem at time t may be constructed by advancing each point of the front in its normal direction a distance t. In terms of our parameterization of the front, the solution is given by

$$x(s,t) = \frac{y_s(s,t=0)}{(x_s^2(s,t=0) + y_s^2(s,t=0))^{1/2}} \, t + x(s,t=0), \qquad (2.8)$$

$$y(s,t) = \frac{-x_s(s,t=0)}{(x_s^2(s,t=0) + y_s^2(s,t=0))^{1/2}} \, t + y(s,t=0). \qquad (2.9)$$

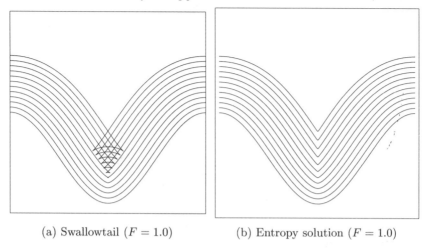

   (a) Swallowtail ($F = 1.0$)   (b) Entropy solution ($F = 1.0$)

Fig. 2.3. Cosine curve propagating with unit speed.

As can be seen in Figure 2.3, the front develops a sharp corner in finite time. Once this corner develops, the normal is ambiguously defined, and it is not clear how to continue the evolution. Thus, beyond the formation of the discontinuity in the derivative, we will need a *weak solution,* so called because the solution weakly satisfies the definition of differentiability.[3]

How can a solution be continued beyond the formation of a singularity in the curvature corresponding to a corner in the front? A reasonable answer depends on the nature of the interface under discussion. If the interface is viewed as a geometric curve evolving under the prescribed speed function, then one possible weak solution is the "swallowtail" solution formed by letting the front pass through itself; this solution is in Figure 2.3(a). This solution is in fact the one given by Eqns. 2.8 and 2.9; the lack of differentiability at the center point does not destroy the solution, since the exact solution is written only in terms of the initial data.

---

[3] A solution is said to be a "weak solution" of a differential equation if it satisfies an integral formulation of the equation. The advantage of such a formulation is that it may not require the same degree of differentiability of a potential solution, and thus may allow more general solutions. As an example, consider the one-dimensional wave equation $u_t = u_x$. A solution to this equation must be differentiable in both x and t. However, if we integrate both sides of the equation with respect to x over the interval $[a, b]$, we then obtain $\frac{d}{dt} \int u\, dx = u(b) - u(a)$, which does not require that the solution be differentiable in space. Weak solutions will be discussed in more detail in Chapter 5.

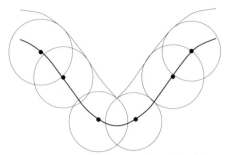

Globally closest points to boundary data

Fig. 2.4. Huygens' solution to propagation $F = 1$.

However, suppose the moving curve is regarded as an interface separating two regions. From a geometrical argument, the front at time t should consist of only the set of all points located a distance t from the initial curve. Figure 2.3(b) shows this alternate weak solution. Roughly speaking, we want to remove the "tail" from the "swallowtail"; see [225]. One way to build this solution is through a Huygens' principle construction; the solution is developed by imagining wave fronts emanating with unit speed from each point of the boundary data and the envelope of these wave fronts always corresponds to the "first arrivals". This will automatically produce the solution given on the right in Figure 2.3. This is the approach taken in [225].

We note that there is a vertical ridge along which two points on the boundary curve are the same distance away, as shown in Figure 2.3b. Along this ridge, the solution is non-differentiable, and the gradient is not defined.

Both of these constructions can be viewed as solutions to the problem of a front propagating with unit speed. However, the solution that we want, corresponding to the shortest distance or "first arrival," is the one obtained through the Huygens' construction. Another way to obtain the solution is through the notion of an *entropy condition* posed by Sethian in [222, 225]; if we imagine the boundary curve as a source for a propagating flame, then the expanding flame satisfies the requirement that once a point in the domain is ignited by the expanding front, it stays burnt. This construction yields the entropy-satisfying Huygens' construction given in Figure 2.4.

What does this "entropy condition" have to do with the notion of "entropy"? While the answer will be made more precise in Chapter 5,

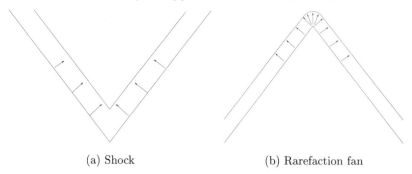

(a) Shock   (b) Rarefaction fan

Fig. 2.5. Front propagating with unit normal speed.

an intuitive answer is as follows. "Entropy" refers to the organization of information. In general terms, an entropy condition is one that says that no new information can be created during the evolution of the problem. Our example shows that once the entropy condition is invoked, some information about the initial data is lost. Indeed, the entropy condition "once a particle is burnt, it stays burnt" means that once a corner has developed, the solution is no longer reversible. The problem cannot be run "backward" in time; if we try to do so, the initial data will not be retrieved. Thus, some information about the solution is forever lost.

As further illustration, consider the case of a V-shaped front propagating normal to itself with unit speed ($F = 1$). In Figure 2.5(a), the point of the front is downward; as the front moves inward with unit speed, a "shock" propagates upward as the front pinches off, and a rule is required to select the correct solution to stop the solution from being multiple-valued. Conversely, in Figure 2.5(b), the point of the front is upward; in this case the unit normal speed results in a circular fan that connects the left state with slope $+1$ to the right state, which has slope -1.

It is important to summarize a key point in this discussion. The choice of weak solution given by our entropy condition[4] rests on the perspective that the front separates two regions and the assumption that one is interested in tracking the progress of one region into the other. Considerable confusion about the level set perspective has resulted from a misunderstanding of the basic assumption inherent in this model.

---

[4] Strictly speaking, this notion of an entropy condition will have meaning only for a propagating graph; a more precise view will be given in Chapter 3. Nonetheless, we shall be somewhat loose with our use of the word "entropy."

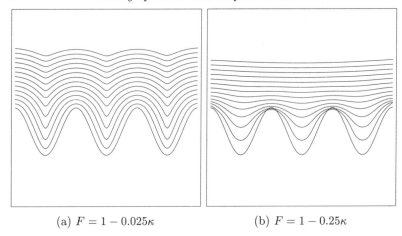

(a) $F = 1 - 0.025\kappa$ (b) $F = 1 - 0.25\kappa$

Fig. 2.6. Propagating triple sine curve.

2.4 Effects of curvature: the viscous limit and the link to hyperbolic conservation laws

Consider now a speed function of the form $F = 1 - \epsilon\kappa$, where ϵ is a constant. The modifying effects of the term $\epsilon\kappa$ are profound, and in fact pave the way toward constructing accurate numerical schemes that adhere to the correct entropy condition.

Following Sethian [225], the curvature evolution equation given by Eqn. 2.3 can be rewritten as

$$\kappa_t = \epsilon\kappa_{\alpha\alpha} + \epsilon\kappa^3 - \kappa^2, \qquad (2.10)$$

where the second derivative of the curvature κ is taken with respect to arc length α. This is a reaction-diffusion equation; the drive toward singularities due to the reaction term $(\epsilon\kappa^3 - \kappa^2)$ is balanced by the smoothing effect of the diffusion term $(\epsilon\kappa_{\alpha\alpha})$.

Consider again the cosine front given in Eqn. 2.7 and the speed function $F(\kappa) = 1 - \epsilon\kappa$, $\epsilon > 0$. As the front moves, the trough at $s = n + 1/2$ is sharpened by the negative reaction term (because $\kappa < 0$ at such points) and smoothed by the positive diffusion term. For $\epsilon > 0$, it can be shown (see [225, 187]) that the moving front stays C^∞. Figure 2.6 shows two cases of a propagating initial triple sine curve. For ϵ small (Figure 2.6(a)), the troughs sharpen up. For ϵ large (Figure 2.6(b)), parts of the boundary with high values of positive curvature initially move downward, and concave parts of the front move quickly up.

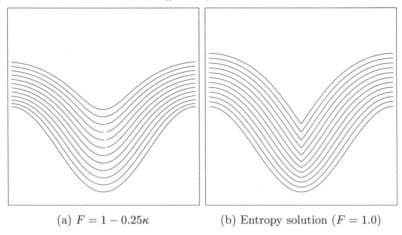

(a) $F = 1 - 0.25\kappa$ (b) Entropy solution $(F = 1.0)$

Fig. 2.7. Entropy solution is the limit of viscous solutions.

For, $\epsilon > 0$, we have a smooth flow, as shown in Figure 2.7(a). However, with $\epsilon = 0$, we have a pure reaction equation $\kappa_t = -\kappa^2$, and the developing corner can be seen in the exact solution $\kappa(s,t) = \kappa(s,0)/(1 + t\kappa(s,0))$. This is singular in finite t if the initial curvature is anywhere negative. The entropy solution to this problem when $F = 1$ is shown in Figure 2.7(b).

The central observation, key to the notion of entropy solutions in front propagation (see [225]), is the following link:

Consider the propagating cosine curve and the two solutions

- $X^\epsilon_{\text{curvature}}(t)$, *obtained by evolving the initial front with* $F_\epsilon = 1 - \epsilon\kappa$,
- $X_{\text{constant}}(t)$, *obtained with speed function* $F = 1$ *and the entropy condition.*

Then, at any time T,

$$\lim_{\epsilon \to 0} X^\epsilon_{\text{curvature}}(T) = X_{\text{constant}}(T). \tag{2.11}$$

Thus, the limit of motion with curvature, known as the "viscous limit", is the entropy solution for the constant speed case.

Why is this known as the viscous limit, and what does this have to do with viscosity? To see why this is an appropriate name, we turn to the link between propagating fronts and hyperbolic conservation laws. The following material, taken from [225], is presented in considerably more depth in Chapter 5. The ideas are presented here as motivation.

An equation for $u(x,t)$ of the form

$$u_t + [G(u)]_x = 0, \qquad (2.12)$$

is known as a "hyperbolic conservation law". A simple example is Burgers' equation, given by

$$u_t + uu_x = 0, \qquad (2.13)$$

which describes the motion of a compressible fluid in one dimension. The solution to this equation can develop discontinuities, known as "shocks", where the fluid undergoes a sudden expansion or compression. These shocks (for example, a sonic boom) can arise from arbitrarily smooth initial data; they are functions of the equation itself. Fluid viscosity appears as a diffusive term on the right-hand side, namely,

$$u_t + uu_x = \epsilon u_{xx}, \qquad (2.14)$$

and this second derivative acts like a smoothing term and stops the development of such shocks. For $\epsilon > 0$ it can be shown that the solution must remain smooth for all time.

What does this have to do with our propagating front equation? Consider the initial front given by the graph of $f(x)$, with f and f' periodic on $[0,1]$, and suppose that the propagating front remains a graph for all time. Let ψ be the height of the propagating function at time t, and thus $\psi(x,0) = f(x)$. The tangent at (x,ψ) is $(1,\psi_x)$. Referring to Figure 2.8, the change in height V in a unit time is related to the speed F in the normal direction by

$$\frac{V}{F} = \frac{(1+\psi_x^2)^{1/2}}{1}, \qquad (2.15)$$

and thus the equation of motion becomes

$$\psi_t = F(1+\psi_x^2)^{1/2}. \qquad (2.16)$$

Use of the speed function $F(\kappa) = 1 - \epsilon\kappa$ and the formula $\kappa = -\psi_{xx}/(1+\psi_x^2)^{3/2}$ yields

$$\psi_t - (1+\psi_x^2)^{1/2} = \epsilon\frac{\psi_{xx}}{1+\psi_x^2}. \qquad (2.17)$$

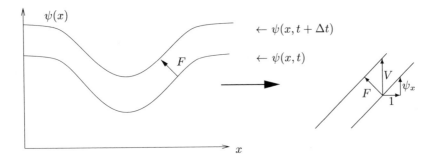

Fig. 2.8. Variables for propagating graph.

This is a partial differential equation with a first order time and space derivative on the left side, and a second order term on the right. Differentiation of both sides of this equation yields an evolution equation for the slope $u = d\psi/dx$ of the propagating front, namely,

$$u_t + [-(1 + u^2)^{1/2}]_x = \epsilon \left[\frac{u_x}{1 + u^2} \right]_x . \qquad (2.18)$$

Thus, as shown by Sethian [226], the derivative of the curvature-modified equation for the changing height ψ looks like some form of a viscous hyperbolic conservation law, similar to Eqn. (2.14), with $G(u) = -(1 + u^2)^{1/2}$ for the propagating slope u. Hyperbolic conservation laws of this form have been studied in considerable detail; in fact, our entropy condition is equivalent to the one for propagating shocks in hyperbolic conservation laws.

To summarize the discussion so far:

(i) A front propagating at a constant speed can form corners as it evolves. At such points, the front is no longer differentiable and a weak solution must be constructed to continue the solution.

(ii) The correct weak solution, motivated by viewing the front as an evolving interface separating two regions, comes by means of an entropy condition.

(iii) A front propagating at a speed $1 - \epsilon\kappa$ for $\epsilon > 0$ does not form corners and stays smooth for all time. Furthermore, as the dependence on curvature vanishes, the limit of this motion is the entropy-satisfying solution obtained for the constant speed case.

(iv) If the propagating curve remains a graph as it moves, there is a direct link between the equation of motion and a one-dimensional hyperbolic conservation law for the evolving slope. The role of curvature in a propagating front is analogous to the role of viscosity in this hyperbolic conservation law.

(v) By considering the initial front as a boundary value, a boundary value partial differential equation can be developed in two and three dimensions for a speed function F which does not change sign. This will lead to the Fast Marching Method.

(vi) By embedding the motion of a hypersurface as the zero level set of a higher dimensional function, an initial value partial differential equation can be obtained that will permit arbitrary speed functions together with curves and surfaces moving in two and three space dimensions. This will lead to the Level Set Method.

Before proceeding to numerical schemes, we first consider a more formal view of the above, which is known as the theory of viscosity solutions.

Viscosity Solutions and Hamilton–Jacobi Equations

Outline: *We present the formal definition of the viscosity solution of a Hamilton-Jacobi equation. This solution is based on its behavior at extrema, which turns out to be a more appropriate way of characterizing the correct weak solution. Using this definition, one then* **proves** *that the viscosity solution is the limit of smooth solutions as the smoothing term goes to zero.*

So far, our presentation of interface propagation has taken a geometric approach. We have tried to present an intuitive feel for the mechanism linking moving fronts and hyperbolic conservation laws. We have done so for two reasons. First, as we have seen, the notion of what happens when a corner develops in an evolving graph neatly parallels the development of shocks and rarefaction fans described by the conservation law for the evolving slope. Second, the rich wealth of numerical schemes developed for hyperbolic equations will be used to build schemes for the initial and boundary value perspectives.

Formally, however, this link cannot be extended to higher dimensions. Recall that for a curve which can be written as a graph propagating with speed F in its normal direction, the change in the height of the function ψ (Eqn. 2.16) is given by

$$\psi_t = F(1 + \psi_x^2)^{1/2}. \tag{3.1}$$

Differentiation of both sides of this equation yields an evolution equation for the slope $u = d\psi/dx$ of the propagating front, namely,

$$u_t + [-(1 + u^2)^{1/2}]_x = \epsilon \left[\frac{u_x}{1 + u^2} \right]_x, \tag{3.2}$$

29

which is a viscous hyperbolic conservation law with $G(u) = -(1+u^2)^{1/2}$ for the propagating slope u.

In contrast, imagine either a level set view that embeds the curve in the higher dimensional level set function ϕ through

$$\phi_t + F(\phi_x^2 + \phi_y^2)^{1/2} = 0, \tag{3.3}$$

or the boundary value view which yields the arrival time function T through

$$|\nabla T|F = 1. \tag{3.4}$$

Suppose we take the first equation and try letting $u = \phi_x$ and $v = \phi_y$. If we then differentiate both sides with respect to x and y, we get a pair of equations for u and v. These are linked by the equality of mixed partials in which $u_y = v_x$. Thus, unlike our formulation for a propagating curve, we cannot simply take the theory for hyperbolic conservation laws and "integrate it upward".

Instead, we work *directly* with the partial differential equations and add a viscous right-hand side. Once again, the solution to this equation is smooth for all time, and the limit as the viscosity term goes to zero produces the appropriate weak solution. In fact, the preferred alternative proceeds along a different line. First, the "viscosity solution" which allows corners is defined in terms of its behavior at extrema, and is shown to be equivalent to the classical (smooth) solution where the solution is smooth. Then, under certain restrictions, it is shown that this viscosity solution is equal to the limit as the viscous smoothing term vanishes. This is the theory of viscosity solutions of Hamilton–Jacobi equations introduced by Crandall and Lions [74]; see also Crandall, Evans, and Lions [71] and Crandall, Ishii, and Lions [72]. In the rest of this brief chapter, we shall make these ideas somewhat more precise; for further details, see Evans [86].

3.1 Viscosity solutions of Hamilton–Jacobi equations

Consider either the level set equation $\phi_t + F|\nabla\phi| = 0$ or the stationary equation $|\nabla T|F = 1$. If the speed F depends only on position x and first derivatives of ϕ, these are particular cases of the more general Hamilton–

Jacobi equation[1]

$$\alpha u_t + H(Du, x) = 0, \qquad (3.5)$$

where $H(Du, x) = F|\nabla u| - (1 - \alpha)$, and α is either zero or one. Here, Du represents the partials of u in each variable, for example, u_x and u_y. Assume that the Hamiltonian H is a smooth function of its arguments. We want to admit non-smooth solutions that allow corners, similar to the previous desire to admit non-smooth solutions of a hyperbolic conservation law. A natural approach, in parallel with the earlier discussion, is to add a viscosity term, that is,

$$\alpha u_t + H(Du, x) = \epsilon \Delta u, \qquad (3.6)$$

where ϵ is a positive constant. Then, given a solution u_ϵ to the foregoing, we want to show that such a solution is smooth, and that the limit of these solutions as ϵ vanishes gives an appropriate weak solution.

Rather than define the weak solution as a limit of smooth solutions, Crandall, Evans, and Lions [71], reformulating an earlier definition by Crandall and Lions [74], define a weak solution as follows:

> **Definition:** A function u is said to be a *viscosity solution* of Eqn. (1.10), if, for all smooth test functions v,
>
> (i) if $u - v$ has a local maximum at a point (x_o, t_o), then
>
> $$v_t(x_o, t_o) + H(Dv(x_o, t_o), x_o) \leq 0 \qquad (3.7)$$
>
> (ii) if $u - v$ has a local minimum at a point (x_o, t_o), then
>
> $$v_t(x_o, t_o) + H(Dv(x_o, t_o), x_o) \geq 0. \qquad (3.8)$$

Note that nowhere in this definition is the viscosity solution u differentiated; everything is done in terms of the test function v. This is done so that one can use the usual trick of integration by parts and move all the derivatives onto the test function in exchange for some boundary conditions.

This is only a definition. Several things need to be checked before it can be viewed as a reasonable solution to the Hamilton–Jacobi equation. In fact, the following can be shown:

---

[1] We are going to ignore curvature-driven fronts for a few moments, because such flows contain second derivatives.

- *If u is a smooth solution of the Hamilton–Jacobi equation, then it is a viscosity solution.*

  In other words, any classical solution that stays smooth for all time satisfies the two inequalities in the above definition.

- *If a viscosity solution u is differentiable at some point, then it satisfies the Hamilton–Jacobi equation at that point.*

  In other words, where the viscosity solution is smooth, it gives the same answer as the classical solution.

- *This viscosity solution is unique, given appropriate initial conditions.*

  That is, there is only one viscosity solution satisfying the definition.

- *The solution produced by taking the limit of the smooth solutions u_ϵ as ϵ goes to zero is the viscosity solution.*

We shall prove none of these here. Precise statements and proofs may be found in [74, 71, 86]. The salient point is that the viscosity solution is now defined in a way that does not require differentiation, and can be proven to be the unique viscous limit of the smoothed Hamilton–Jacobi equation.

3.2 Some additional comments and references

While methods for numerically approximating moving fronts have received much attention, the theoretical analysis of moving curves and surfaces has itself been a subject of considerable interest. The work of Gage [97], Gage and Hamilton [98], and Grayson [105], discussed later in this book, opened up the analysis of curve evolution under curvature, leading to the beautiful result that a closed curve shrinking under its curvature collapses smoothly to a point.

Using a very different approach, Brakke [35] applied varifold theory to the problem of a hypersurface moving under its curvature, and in so doing provided a wide-ranging perspective for these problems, including cases in which the evolving curves were not necessarily smooth.

There has been considerable theoretical analysis of the Eulerian partial differential equations view of interface motion and its relation to other perspectives on front propagation. The flame/entropy model from [222] served as the basis for theoretical analysis by Barles [19]. The embedding of the front as a higher dimensional function meant that some of the issues of topological change and corner formation could be studied in a more natural manner. Furthermore, the transformation of geometry problems into a partial differential equation setting meant that some powerful analytic techniques, including regularity of solutions, viscous solutions of Hamilton–Jacobi equations, and tools for analyzing existence and uniqueness, could be applied.

Using our perspective developed in [225, 187], Evans and Spruck [88, 89, 90, 91] and Chen, Giga, Goto, and Ishii [57, 101, 102] performed detailed analysis of curvature flow in a series of papers. They exploited much of the work on viscosity solutions of partial differential equations (see Lions [153]), which itself was inspired by the corresponding work applied to hyperbolic conservation laws. Their papers examined the regularity of curvature flow equations, pathological cases, and the link between the level set perspective and the varifold approach of Brakke. These papers opened up a series of investigations into further issues; we also refer the interested reader to Evans, Soner, and Souganidis [87], and Ilmanen [119, 120].

For the stationary boundary value view, the interested reader is referred to a large body of literature, including relevant theory by Barles and Souganidis [22], Crandall, Evans, and Lions [71], Lions [153], and Souganidis [253], and numerical algorithms by Bardi and Falcone [18], Falcone, Giorgi, and Loretti [93], Kimmel and Bruckstein [130], and Rouy and Tourin [207].

Finally, there has also been considerable work on the analysis of relevant numerical algorithms for viscosity solutions outside of the context of interface propagation. Crandall and Lions [74] analyzed an explicit finite difference scheme; see also, for example, Souganidis [253]. Most of these schemes are finite difference expressions that contain some forms of smoothing, often in the spirit of the method of artificial viscosity. Convergent schemes result from the standard criteria of monotonicity and consistency.

4
Traditional Techniques for Tracking Interfaces

Outline: *Before focusing on the initial and boundary value partial differential equations view of evolving fronts, we consider other numerical methods for tracking interfaces. We then make a first attempt at solving a partial differential equation for front propagation using a central difference scheme, and show why this fails to capture the correct weak solution.*

4.1 Marker/string methods

A standard approach to modeling moving fronts comes from discretizing the Lagrangian form of the equations of motion given in Eqn. (2.1). In this approach, the parameterization is discretized into a set of marker particles whose positions at any time are used to reconstruct the front. This approach is known under a variety of names, including marker particle techniques, string methods, and nodal methods. In two dimensions, the front may be reconstructed as line segments; in three dimensions, triangles are often used.

This approach can be illustrated through a straightforward scheme that constructs a simple difference approximation to the Lagrangian equations of motion. Divide the parameterization interval $[0, S]$ into M equal intervals of size Δs, yielding $M + 1$ mesh points $s_i = i\Delta s$, $i = 0, \ldots, M$, as shown in Figure 4.1. Divide time into equal intervals of length Δt. The image of each mesh point $i\Delta s$ at each time step $n\Delta t$ is a marker point (x_i^n, y_i^n) on the moving front. The goal is a *numerical algorithm* that will produce new values (x_i^{n+1}, y_i^{n+1}) from the previous positions; we follow the discussion in [225].

First, we approximate parameter derivatives at each marker point by using neighboring mesh points. Central difference approximations based

34

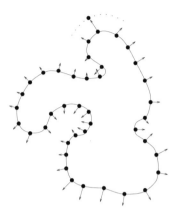

Fig. 4.1. Discrete parameterization of curve.

on Taylor series (see Chapter 5) yield

$$\frac{dx_i^n}{ds} \approx \frac{x_{i+1}^n - x_{i-1}^n}{2\Delta s}, \qquad \frac{dy_i^n}{ds} \approx \frac{y_{i+1}^n - y_{i-1}^n}{2\Delta s}, \qquad (4.1)$$

$$\frac{d^2 x_i^n}{ds^2} \approx \frac{x_{i+1}^n - 2x_i^n + x_{i-1}^n}{\Delta s^2}, \qquad \frac{d^2 y_i^n}{ds^2} \approx \frac{y_{i+1}^n - 2y_i^n + y_{i-1}^n}{\Delta s^2}. \qquad (4.2)$$

Similarly, time derivatives may be replaced by the forward difference approximations

$$\frac{dx_i^n}{dt} \approx \frac{x_i^{n+1} - x_i^n}{\Delta t}, \qquad \frac{dy_i^n}{dt} \approx \frac{y_i^{n+1} - y_i^n}{\Delta t}. \qquad (4.3)$$

Substitution of these approximations into the equations of motion given by Eqn. 2.1 produces the scheme

$$(x_i^{n+1}, y_i^{n+1}) = (x_i^n, y_i^n) + \Delta t \, F(\kappa_i^n) \frac{(y_{i+1}^n - y_{i-1}^n, -(x_{i+1}^n - x_{i-1}^n))}{((x_{i+1}^n - x_{i-1}^n)^2 + (y_{i+1}^n - y_{i-1}^n)^2)^{1/2}}, \qquad (4.4)$$

where

$$\kappa_i^n = 4 \frac{(y_{i+1}^n - 2y_i^n + y_{i-1}^n)(x_{i+1}^n - x_{i-1}^n) - (x_{i+1}^n - 2x_i^n + x_{i-1}^n)(y_{i+1}^n - y_{i-1}^n)}{((x_{i+1}^n - x_{i-1}^n)^2 + (y_{i+1}^n - y_{i-1}^n)^2)^{3/2}}. \qquad (4.5)$$

Using the fact that the curve is closed, this is a complete recipe for updating the positions of the particles from one time step to the next.

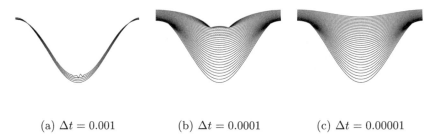

(a) $\Delta t = 0.001$ (b) $\Delta t = 0.0001$ (c) $\Delta t = 0.00001$

Fig. 4.2. Marker particle solution to $F = 1 - .25\kappa$.

Observe that the fixed discretization interval Δs has dropped out of the expression. Consequently, as marker particles come together, quotients on the right-hand side of Eqn. 4.4 approach zero over zero, yielding a very sensitive calculation. The computed curvature can change drastically from one particle to the next because of small and unavoidable errors in the positions.

We can observe this unstable accumulation of small errors by using a marker particle scheme to follow our previous initial cosine curve propagating with speed $1 - \epsilon\kappa$, $\epsilon = .25$. Since $\epsilon > 0$, the exact solution is always smooth and differentiable. Fifty marker points are used, together with a time step of $\Delta t = 0.001$. Although the propagating front begins to sharpen as expected (see Figure 4.2(a)), oscillations soon develop, which grow uncontrollably. These oscillations result from a feedback cycle: (1) small errors in approximate marker positions produce (2) local variations in the computed derivatives leading to (3) variation in the computed particle velocities causing (4) uneven advancement of markers, which yields (5) larger errors in approximate marker positions. Within a few time steps, the small oscillations in the curvature have grown and the computed solution becomes unbounded. Figure 4.2(a) shows the calculation until the computer program stops running.

Suppose we try to increase accuracy by using a smaller time step. Figure 4.2(b) and Figure 4.2(c) show calculations with $\Delta t = 0.0001$ and $\Delta t = 0.00001$, respectively. With time step $\Delta t = .0001$, once again the solution becomes unstable, and the smooth decay of the trough is not seen. Only in the finest case using a time step of $\Delta t = 0.00001$ is the solution acceptable. For any $\epsilon > 0$ there is a bound on the minimum distance between particles, and thus a small enough time step does exist to ensure stability. However, note that if we had chosen a smaller value of ϵ, the marker trajectories would have come closer together, and a smaller time step would be required for stability. As an example, with the value of $\epsilon = 0.1$ (a large value, when one considers the role of curvature to be similar to that of surface tension), the time step required for stability is $\Delta t = 0.000005$. For values of ϵ much smaller, this approach becomes impractical.

What can be done? Typically, there are three remedies:

- "Smooth" the speed function so that the marker points stay far enough apart to allow a reasonable time step.
- Redistribute marker particles according to arc length (or a related quantity) every few time steps so that they stay far enough apart.
- Invent some filtering technique to remove "noise" (oscillations) in the particle positions as they develop.

While all three techniques are used in practice, they are not appealing. They all boil down to the same thing; they alter the equations of motion in non-obvious ways. Significant amounts of smoothing may be required to ensure a practical time step. Thus, one sacrifices the most interesting propagation characteristics, such as front sharpening and curvature singularities, simply in order to keep the calculation alive. Calculation of arc length to redistribute the particles adds a additional smoothing term to the speed function that is difficult to analyze. The computed solution may be far from the desired one. In the worst case, time and effort will be spent solving an unrelated problem.

The situation is even bleaker in the limiting case $\epsilon = 0$. As discussed earlier, a solution forms a sharp corner, and an entropy condition must be invoked to produce a reasonable weak solution beyond the formation of the singularity. However, a marker particle approach does not "know" about the necessary entropy condition, because it attempts to track a Lagrangian formulation in which the swallowtail solution given in Figure 2.3(a) is an acceptable weak solution. In Figure 4.3 we show a marker particle solution that incorporates the swallowtail solution.

No time step, no matter how small, can produce a scheme which correctly incorporates the entropy condition given in [222]. In fact, for $\epsilon > 0$, the equations for the markers themselves reduce to a linearly unstable hyperbolic system (see [187]). From an algorithmic point of view, markers must somehow be eliminated from the discretization as corners form and information about the initial data is lost. This corresponds to deleting the "tail" from the swallowtail, as discussed earlier;

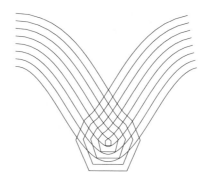

Fig. 4.3. Marker particle solution to swallowtail under $F = 1$.

in some arenas, this procedure is called "de-looping". While there have been procedures for doing so in two dimensions [112], de-looping in three dimensions is not for the fainthearted, and such codes can break down in complex situations.

These drawbacks to marker particle/Lagrangian methods result from stability and local singularity problems. Topological changes in the moving front are also problematic. Consider two separate expanding flames. Suppose these fronts merge and the boundary becomes a single curve. It is difficult to produce a systematic way of removing those markers that no longer sit on the actual boundary. The bookkeeping of removing, redistributing, and connecting markers is complex, and an arduous task for higher-dimensional interface problems.

To summarize, Lagrangian approximations provide numerical schemes based on a parameterized description of the moving front. They can be accurate for small-scale motions of interfaces. However, because they follow a local representation of the front rather than using a global one that is able to take into account the proper entropy conditions and weak solutions, they can suffer from instability and topological limitations.

4.2 Volume-of-fluid techniques

A different approach to front motion is provided by volume-of-fluid techniques, introduced by Noh and Woodward [183] and based instead on an Eulerian view. These techniques have appeared in a variety of forms and

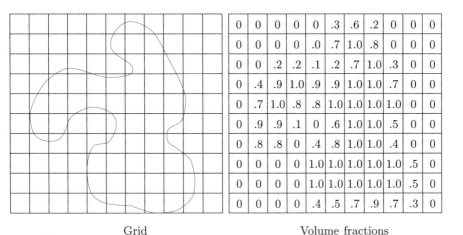

0	0	0	0	0	.3	.6	.2	0	0	0
0	0	0	0	.0	.7	1.0	.8	0	0	0
0	0	.2	.2	.1	.2	.7	1.0	.3	0	0
0	.4	.9	1.0	.9	.9	1.0	1.0	.7	0	0
0	.7	1.0	.8	.8	1.0	1.0	1.0	1.0	0	0
0	.9	.9	.1	0	.6	1.0	1.0	.5	0	0
0	.8	.8	0	.4	.8	1.0	1.0	.4	0	0
0	0	0	0	1.0	1.0	1.0	1.0	1.0	.5	0
0	0	0	0	1.0	1.0	1.0	1.0	1.0	.5	0
0	0	0	0	.4	.5	.7	.9	.7	.3	0

Grid   Volume fractions

Fig. 4.4. Volume-of-fluid method.

have been reintroduced under many names, such as the "cell method" and the "method of partial fractions."

The basic idea (see Figure 4.4) is as follows. Imagine a fixed grid on the computational domain, and assign values to each grid cell based on the fraction of that cell containing material inside the interface. Given a closed curve, we assign a value of unity to those cells completely inside this curve, a cell value of zero to those completely outside, and a fraction between 0 and 1 to cells that straddle the interface, based on the amount of the cell inside the front.

The idea, then, is to rely solely on these "cell fractions," shown in Figure 4.4, to characterize the interface location. Approximation techniques are then used to reconstruct the front from these cell fractions. The original Noh and Woodward algorithm was known as "SLIC", for "Simple Line Interface Calculation," and reconstructed the front as either a vertical line or a horizontal line.

In order to evolve the interface, the cell fractions on this fixed grid are updated to reflect the progress of the front. Suppose that one wishes to advect a front passively under the transport velocity $\vec{u}$ (here, this is not a speed normal to the front, but merely a transport term). Noh and Woodward provide a methodology in which the value in each cell is updated under this transport velocity in each coordinate direction by locally reconstructing the front and then exchanging material in neigh-

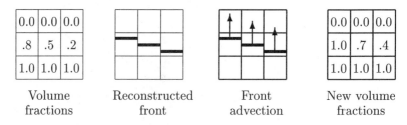

| Volume fractions | Reconstructed front | Front advection | New volume fractions |

Fig. 4.5. Reconstruction and advection of volume fractions.

boring cells under this motion. After completing coordinate sweeps, one has produced new cell fractions at the next time step corresponding to the updated front. In Figure 4.5, we show the advection of an interface under a simple vertical velocity field $\vec{u} = (0, 1)$.

Since the introduction of volume-of-fluids methods, many elaborate reconstruction techniques have been developed to include pitched slopes and curved surfaces, see Chorin [65], Hirt and Nicholls [113], and Lafaurie, Nardone, Scardovelli, Zaleski, and Zanetti [142]. The resulting accuracy depends on the sophistication of the reconstruction and the "advection sweeps" which advance the material. Some of the most elaborate and accurate versions of these schemes to date are due to Puckett [198].

The original SLIC algorithm was designed for transport under an advection velocity field that depended on the location of the front, but not the local shape/orientation of the front. A valuable extension to volume-of-fluid techniques was developed by Chorin [65], who provided a way of using these methods for a speed function given in a direction normal to the front. As we have seen earlier, corners and cusps can develop, and an entropy condition must be invoked to provide the correct weak solution beyond the occurrence of singularities. Chorin did so by relying on a Huygens' principle construction. He considered each spot on the front as a point source, and approximated the envelope of the influence of these sources to give the new position of the front. By advancing the front in enough directions to approximate this point source, this construction satisfies the entropy condition and provides the correct weak solution. This version of SLIC was used in a collection of flame propagation/combustion calculations; see Chorin [65], Ghoniem, Chorin, and Oppenheim [100], and Sethian [224].

It is the Eulerian nature of volume-of-fluid techniques that allows them to avoid many of the Lagrangian time step and topological change problems that plague marker particle methods. As such, they can be quite useful.

However, there are some drawbacks:

- Such techniques are inaccurate. Since the approximation to the front through volume fractions is relatively crude, a large number of cells are often required to obtain reasonable results.
- Evolution under complex speed functions is problematic. The results are noticeably dependent on the underlying orientation of the grid). These problems become worse in the presence of directional velocity fields, such as those arising from non-convex laws.
- Calculation of intrinsic geometric properties of the front, such as curvature and normal direction, may be inaccurate.
- Considerable work may be required to develop higher order versions of such schemes. In higher dimensions, accurate calculations of mean and Gaussian curvatures, especially at saddle points, are difficult to perform.

Nonetheless, such schemes can be powerful, and we refer the interested reader to Puckett [198]. [1]

4.3 A first attempt at constructing an approximation to the gradient

We now turn to our previous partial differential equations (PDE) view of evolving interfaces, and attempt to construct a numerical approximation. The marker particle method discretizes the front. The volume-of-fluid (VOF) method divides the domain space into cells that contain fractions of material. Our Eulerian PDE approach divides the domain into grid points that hold approximations to either the level set function $\phi(x, y, z, t)$ or the arrival time function $T(x, y, z)$. Thus, we need accurate approximations to the gradients $\nabla \phi$ and ∇T.

---

[1] In fact, the author's interest in front propagation schemes stemmed from Chorin's 1980 Huygens' principle construction [65]. Chorin use of discrete, cell-based front algorithms to mimic this expanding wavefront construction, while somewhat crude, was the first to try viewing motion in a normal direction as the Huygens' expansion.

An attempt by the author, who was Chorin's Ph.D. student, to try this technique in three dimensions led to his 1982 doctoral dissertation [222] analyzing curve and surface evolution and developing entropy conditions for front propagation in the context of a flame/entropy model, the 1985 paper [225] on the role of curvature as viscosity and connections of front motion with hyperbolic conservation laws, and the 1987 paper [226] proposing using upwind shock schemes for tracking interfaces. These led to Level Set Methods [187] and Fast Marching Methods [233].

To motivate these approximations, we turn to the simpler case of an evolving curve whose position can always be described as the graph of a function. The equation of motion for the height function $\psi(x,t)$ was given in Eqn. 2.16 by

$$\psi_t = F(1 + \psi_x^2)^{1/2}. \tag{4.6}$$

Just as was done in the Lagrangian case, one might try a scheme which replaces all spatial derivatives with central differences and the time derivative with a forward difference. However, such an algorithm may not work. Let $F(\kappa) = 1$ and consider the initial value problem

$$\psi_t = (1 + \psi_x^2)^{1/2}, \tag{4.7}$$

$$\psi(x,0) = f(x) = \left\{ \begin{array}{ll} 1/2 - x & x \le 1/2 \\ x - 1/2 & x > 1/2 \end{array} \right\}. \tag{4.8}$$

The initial front is a "V" formed by rays meeting at $(1/2, 0)$. Invoking the Huygens' construction, the solution at any time t is the set of all points located a distance t from the initial "V." To build a central difference numerical scheme, we mesh the interval $[0, 1]$ into $2M - 1$ points and form the approximation to the spatial derivative ψ_x in Eqn. 4.7 given by[2]

$$\psi_t \approx \frac{\psi_i^{n+1} - \psi_i^n}{\Delta t} = \left[1 + \left[\frac{\psi_{i+1}^n - \psi_{i-1}^n}{2\Delta x} \right]^2 \right]^{1/2}. \tag{4.9}$$

For all $x \ne 1/2$, ψ_t is correctly calculated to be $\sqrt{2}$, since the graph is linear on either side of the corner and thus the central difference approximation is exact. However, since $x_M = 1/2$, by symmetry, $\psi_{M+1} = \psi_{M-1}$, thus the right-hand side is 1 at x_M. This has nothing to do with the size of the space step Δx or the time step Δt. *As long as we use an odd number of evenly spaced number of points the approximation to ψ_t at $x = 1/2$ gets no better, no matter how small we take the numerical parameters.* It is simply due to the way in which the derivative ψ_x is approximated. Figure 4.6 shows results using this scheme, with the time derivative ψ_t replaced by a forward difference scheme.

It is easy to see what has gone wrong. In the exact solution, $\psi_t = \sqrt{2}$ for all $x \ne 1/2$. This should also hold at $x = 1/2$ where the slope ψ_x is not defined; the Huygens' construction sets $\psi_t(x = 1/2, t)$ equal

---

[2] We postpone a discussion of finite difference approximations until Chapter 5.

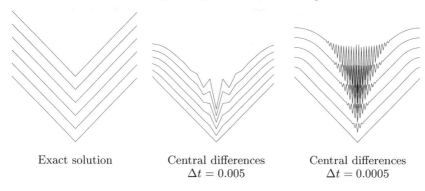

| Exact solution | Central differences $\Delta t = 0.005$ | Central differences $\Delta t = 0.0005$ |

Fig. 4.6. Central difference approximation to level set equation.

to $\lim_{x \to 1/2} \psi_t$. Unfortunately, the central difference approximation chooses a different (and, for our purpose, wrong) limiting solution. It sets the undefined slope ψ_x equal to the average of the left and right slopes. As the calculation progresses, this miscalculation of the slope propagates outward from the spike as oscillations. Eventually, these oscillations cause blowup in the code.

It is clear that more care must be taken in formulating an algorithm. We require schemes that can approximate the gradient term $|\nabla \phi|$ in a way that correctly accounts for the corners. This is the topic of the next chapter.

5

Hyperbolic Conservation Laws

Outline: *The aim of this chapter is to develop numerical schemes which both solve the both initial value equation*

$$\phi_t + F|\nabla\phi| = 0$$

and the boundary value equation

$$F|\nabla T| = 1$$

and select the correct weak solution corresponding to the viscous limit of the associated curvature-driven equation. This chapter is a review of the basics of the technology for hyperbolic conservation laws, and it may be skipped if the reader is already acquainted with this material; good reviews may be found in the monograph by Lax [145] and the book by LeVeque [149]. The goal at the end of this chapter is an understanding of basic schemes for approximating a single hyperbolic conservation law. To motivate such schemes, we start with a simple first order, constant coefficient wave equation.

5.1 The linear wave equation

5.1.1 First order schemes

Consider the one-dimensional wave equation

$$u_t(x,t) + u_x(x,t) = 0, \quad \text{with} \quad u(x,0) = f(x). \quad (5.1)$$

The exact solution to this equation is given by $u(x,t) = f(x-t)$, which can be checked by differentiation using the chain rule. This means that the solution u at any point x at time t is the same as the value of the initial data at the point $x - t$ on the x axis. Another way to say this

44

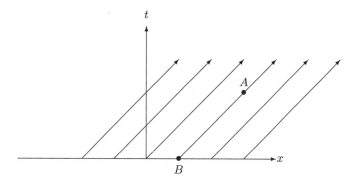

Fig. 5.1. Solution to $u_t = -u_x$ is constant along lines of slope 1.

is that the solution u is constant on lines of slope 1 drawn in the $x - t$ plane (see Figure 5.1).

Consider a point A located in the $x - t$ plane (see Figure 5.1). The solution at point A can be found by tracing back along a line with slope 1 to the point B on the initial line; hence we say that the *domain of dependence* of the point A is the point B. Conversely, the *domain of influence* of point B is the set of all points on the line with slope 1 emanating from B. We now show that these curves in $x - t$ space, which give rise to domains of dependence and influence and are known as *characteristics*, are important in constructing appropriate numerical schemes.

In order to approximate the equation $u_t + u_x = 0$, we begin by following the standard approach and discretize $x - t$ space into a collection of grid points so that $\Delta x = h$ and $\Delta t = k$. Thus every grid point can be represented by the coordinate pair (i, n) corresponding to the point (ih, nk) (see Figure 5.2).

Consider now the various ways of approximating the equation $u_t + u_x = 0$, which we rewrite as $u_t = -u_x$. We begin with the left side of the equation. The solution u at time $t + \Delta t$ can be expanded as a Taylor series in time around the point (x, t); thus we have

$$u(x, t + k) = u(x, t) + u_t(x, t)k + O(k^2), \qquad (5.2)$$

where the expression $O(k^2)$ includes all terms of order k^2 or higher. Rearranging this expression, we can then write the time derivative at

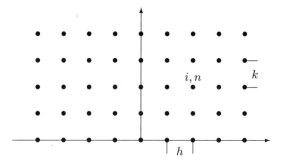

Fig. 5.2. Computational grid.

the point (x, t) as

$$u_t = \frac{u(x, t+k) - u(x, t)}{k} + O(k). \qquad (5.3)$$

This is known as a *forward difference* operator for the time variable, because we have used a Taylor series ahead in time to approximate u_t. Define the notation

$$D^{+t} u \equiv \frac{u(x, t+k) - u(x, t)}{k}. \qquad (5.4)$$

We can then rewrite Eqn. 5.3 as

$$u_t = D^{+t} u + O(k). \qquad (5.5)$$

What about the spatial derivative u_x? Define the operators

$$\begin{aligned}
D^{+x} u &\equiv \frac{u(x+h, t) - u(x, t)}{h}, \\
D^{-x} u &\equiv \frac{u(x, t) - u(x-h, t)}{h}, \\
D^{0x} u &\equiv \frac{u(x+h, t) - u(x-h, t)}{2h}.
\end{aligned} \qquad (5.6)$$

In the same manner, we can construct forward, backward, or centered Taylor series expansions in x for the value u around the point (x, t) can be constructed to produce

$$u_x = D^{+x} u + O(h), \quad u_x = D^{-x} u + O(h), \quad u_x = D^{0x} u + O(h^2). \qquad (5.7)$$

Although the last is a more accurate approximation, we shall see in a later section that accuracy is not the only concern. Nonetheless, these approximations lead to three distinct schemes for computing the solution to the equation $u_t = -u_x$. Let u_i^n be the computed solution at time nk at the point ih. We then have the schemes

(i) *Forward scheme:* $u_i^{n+1} = u_i^n - kD^{+x}u_i^n$.
(ii) *Backward scheme:* $u_i^{n+1} = u_i^n - kD^{-x}u_i^n$.
(iii) *Centered scheme:* $u_i^{n+1} = u_i^n - kD^{0x}u_i^n$.

Which scheme should we use? One view lies in the previous discussion about domains of dependence and characteristics. Recall that the solution u is constant along lines of slope 1 in the $x-t$ plane; this means that information is propagating from the left to the right. Consider now the three difference operators:

- D^{+x} computes the new value at i using information at i and $i+1$; thus information for the solution propagates from right to left.
- D^{-x} computes the new value at i using information at i and $i-1$; thus information for the solution propagates from left to right.
- D^{0x} computes the new value at i using information at $i+1$ and $i-1$; thus information for the solution propagates from both sides.

From this discussion alone, we can dismiss the forward difference scheme from consideration. The backward scheme is referred to as an *upwind scheme*, because it uses values upwind of the direction of information propagation; clearly, this is desirable. Another way to say this is that *"the numerical domain of dependence should contain the mathematical domain of dependence."*

Thus, the backward difference scheme correctly respects the upwind nature of the differential equation and sends information in the direction that correctly matches the differential equation.[1]

5.1.2 Higher order schemes for the linear wave

Can we construct a scheme of higher order space accuracy by trying to knock off more terms in the Taylor series? Using Taylor series and the

---

[1] An associated issue is the "stability" of a scheme, that is, what it does to small errors in the initial data. A more precise way to analyze the stability is through a "Fourier stability analysis," which is performed by considering a set of discrete grid data as wave numbers of the discrete Fourier transform of some periodic function. Then the differential operators in the scheme are equivalent to algebraic operations on the data, and an "amplification factor" can be derived that measures how energy in a particular wave number (that is, data) is amplified from one time step to the next. Stability is examined by finding those values for the time step and space step such that the amplification factor is less than unity; if such a condition is satisfied, then small errors in the solution cannot grow uncontrollably. Typically, the stability of a scheme depends on a balance between the time step k, the space step h, and the speed of the equation; this is known as the Courant-Friedrichs-Levy condition. For our constant unit speed wave equation, stability for the backward scheme will require that $k/h \leq 1$.

equation $u_t = -u_x$, we have that

$$u(x, t+k) = u + ku_t + \frac{k^2}{2}u_{tt} + O(k^3) = u - ku_x + \frac{k^2}{2}u_{xx} + O(k^3). \quad (5.8)$$

Thus, this suggests the scheme

$$u_i^{n+1} = u_i^n - kD^{0x}u_i^n + \frac{k^2}{2}D^{+x-x}u_i^n. \quad (5.9)$$

(Here, the notation $D^{+x-x}u_i^n$ is the centered approximation to the second derivative given by $D^{+x-x}u \equiv \frac{u(x+h,t)-2u(x,t)+u(x-h,t)}{h^2}$.) This scheme is known as *Lax–Wendroff*, and is second order accurate.

The last term in the Lax–Wendroff scheme looks like a diffusion term. We can think of this scheme as approximating the solution to the advection-diffusion equation

$$u_t = -u_x + \frac{k}{2}u_{xx}, \quad (5.10)$$

where the size of the smoothing second derivative term depends on the time step. This is very closely related to the discussion of viscous limits in Chapters 2 and 3. Lax-Wendroff uses a second order diffusive term to smooth discontinuities in the solution. As the numerical method is refined, the scheme converges to the correct solution to the original problem.

Why not always use a higher order scheme? If the solution is smooth for all time, then the additional accuracy offered by a higher order scheme may be worth it. However, preserving corners may be important, and higher order schemes can smooth them out. In order to build schemes that handle corners correctly, we need to focus on techniques that can treat sharp discontinuities.

5.2 The non-linear wave equation

5.2.1 Discontinuous solutions and shocks

Let's continue and examine a wave equation with a non-constant speed. Consider an equation of the form

$$u_t + a(x)u_x = 0, \quad (5.11)$$

in which the propagation speed depends on a known function $a(x)$ of the position. We can build variations of the previous upwind schemes that will select the correct direction of the "upwinding." The direction depends on the sign of a. For example, the scheme

$$u_i^{n+1} = u_i^n - \Delta t[\max(0, a_i)D^{-x}u_i^n + \min(0, a_i)D^{+x}u_i^n] \qquad (5.12)$$

chooses the correct direction of the upwinding so that the difference scheme always includes the mathematical domain of dependence.[2] Note how this scheme works. When the propagation speed a is positive (as it was in our constant coefficient case), information travels from left to right, and the backward difference operator is selected; when the wave speed a is negative, the forward difference operator is selected. As in the constant speed wave equation, we will need to pick a time step such that $k/h \leq 1/M$, where M is a maximum bound on the wave speed a.

Pushing further, what about the fully non-linear equation, for example,

$$u_t + uu_x = 0, \qquad (5.13)$$

in which the propagation speed depends on the value of u itself? We first observe that the solution u is still constant along lines that leave the initial line $t = 0$ in $x - t$ space. To see that this is true, consider a particle moving through the $x - t$ plane whose position x at any time t is parameterized by s. Then, by the chain rule,

$$\frac{du(x(s), t(s))}{ds} = u_x \frac{dx}{ds} + u_t \frac{dt}{ds} = \frac{dt}{ds}\left[u_t + \frac{dx}{dt}u_x\right]. \qquad (5.14)$$

Now, suppose the trajectory of the particle is set so that $\frac{dx}{dt} = u$. The right-hand side is zero by the differential equation, and thus $\frac{du}{ds} = 0$. Thus, u is constant along the trajectories, known as "characteristics," which means that the slope does not change, hence these characteristics are straight lines. However, the straight lines are no not necessarily parallel, and this gives rise to ambiguity in the evolving equations.

As an example, consider the earlier non-linear equation and initial data

$$u(x, 0) = \left\{ \begin{array}{ll} 1 & x \leq 0 \\ 1 - x & 0 < x < 1 \\ 0 & x \geq 1 \end{array} \right\}. \qquad (5.15)$$

For these data, the characteristics are straight lines along which the solution u is constant and hence transported; they are shown in Figure 5.3.

---

[2] Here, a_i means $a(ih)$.

Shock: 1/slope = shock speed

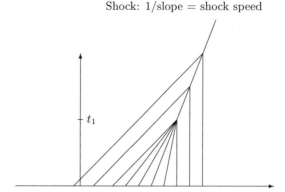

Fig. 5.3. Converging characteristics: formation of shocks.

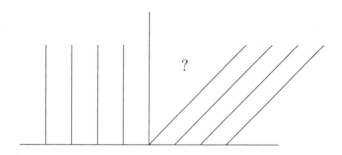

Fig. 5.4. Diverging characteristics: a gap in the solution.

At some finite time t_1, the characteristics collide. Each colliding characteristic leaves the initia line with slope $\frac{1}{u}$ and carries its own value of u. Along the line that marks the collision, known as a shock, the solution discontinuously jumps from the left state (which corresponds to a u value of 1, as seen from the initial data) to the right state of 0. Beyond the collision time it is not clear how to carry the solution ahead uniquely in time.

If the problem is reversed, the issue is equally unclear. Consider now the same equation with initial data

$$u(x,0) = \left\{ \begin{array}{ll} 0 & x < 0 \\ 1 & x \geq 0 \end{array} \right\}. \qquad (5.16)$$

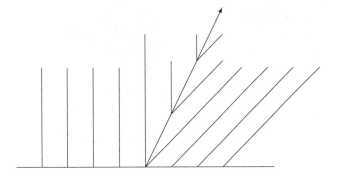

Fig. 5.5. Rarefaction shock.

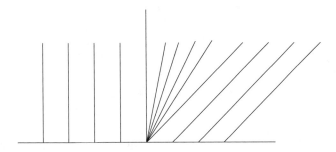

Fig. 5.6. Rarefaction fan.

Again, the characteristics are straight lines transporting the solution; the solution is graphed in Figure 5.4.

The empty gap containing the question mark is called a "rarefaction zone" because of its connection with fluid mechanics; the solution must somehow expand from $u = 0$ on the left to $u = 1$ on the right to fill in the rarefied area. What is the proper way to build the solution in the empty gap? One can think of at least two solutions.

One option is to put in a shock, as shown in Figure 5.5. An alternative is to connect the two states by means of a fan, as in Figure 5.6. Which solution is correct? In order to answer all of these questions, consider the associated *viscous non-linear wave equation*

$$u_t + uu_x = \epsilon u_{xx}. \tag{5.17}$$

This is the same non-linear wave equation with a second derivative added to the right-hand side. This viscous second derivative smooths out sharp corners as they develop. Thus, the solution stays smooth for all time; see Lax [145] and Chorin and Marsden [67]. We thus *choose* our solution to the non-linear wave equation with zero viscosity to be the one obtained as the limit of the solution to the viscous non-linear equation as the viscosity coefficient ϵ vanishes. It can be shown that an alternate way to construct this "entropy solution" comes from enforcing an "entropy condition" which requires that characteristics flow *into* shocks (Figure 5.3), rather than emanate from them (Figure 5.5). Using this definition, the rarefaction fan of Figure 5.6 is the entropy solution. Our goal is to build numerical schemes for the non-linear equation that satisfy this entropy condition.

Before doing so, it is instructive to recall the initial V shapes of Figure 2.5 propagating with constant speed. When the corner pointed inward(that is, the slope on the left was negative and that on the right was positive), a corner formed in the propagating curve, and a shock formed where the normals collided. In the case of an outward pointing corner, where the signs of the slopes were switched, a fan developed as the front moved outwards. Both solutions are limiting solutions of the equation for a front propagating with curvature-dependent speed $(1 - \epsilon\kappa)$ as the curvature coefficient ϵ vanishes. This is analogous to the earlier discussion and is the fundamental reason why we are led to analyzing schemes for shocks and hyperbolic equations.

5.2.2 Weak solutions, flux condition, and approximation schemes

We now discuss numerical schemes that allow physically reasonable and entropy-satisfying non-differentiable solutions, known as "weak solutions." Consider the first order non-linear hyperbolic equation

$$u_t + [G(u)]_x = 0. \tag{5.18}$$

If $G(u) = u$, this gives the linear constant coefficient case; if $G(u) = u^2/2$, this gives the non-linear equation $u_t + uu_x = 0$. Integration of both sides of the equation produces

$$
\begin{aligned}
0 &= \int_a^b (u_t + [G(u)]_x)dx \\
&= \int_a^b u_t dx + \int_a^b [G(u)]_x dx
\end{aligned}
$$

$$= \frac{d}{dt} \int_a^b u \, dx + G(u(b,t)) - G(u(a,t)). \qquad (5.19)$$

Thus,

$$\frac{d}{dt} \int_a^b u \, dx = G(u(a,t)) - G(u(b,t)). \qquad (5.20)$$

This leads to a physical interpretation of the hyperbolic conservation law given in Eqn. 5.18; the change in the amount of u between a and b is equal to the flux $G(u)$ flowing into the interval. $G(u)$ is referred to as the *flux* function (see Figure 5.7).[3]

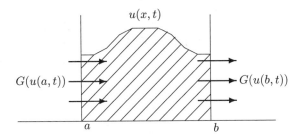

Fig. 5.7. Flux G of substance u into interval $[a, b]$.

We say that u is *conserved* under Eqn. (5.20) because there is a balance between the change of u in the interval $[a, b]$ and the flux of material into the interval. The reason we have re-written Eqn. 5.18 in the conservative form of Eqn. 5.20 is that the latter does not assume that the solution is differentiable with respect to x. Thus, the solution class of equations has been markedly broadened.

Eqn. 5.18 allows us to compute the speed of the shock in Figure 5.3 as follows. Pick three points x_1, x_*, and x_2 such that at time t the shock is at $(x_*(t), t)$ with $x_1(t) < x_*(t) < x_2(t)$. Let u_1 be the constant initial state to the left of the shock, and u_2 the constant initial state to the right of the shock. Then the shock speed $S = \frac{dx_*}{dt}$ can be found by the following:

$$\frac{d}{dt} \int_{x_1}^{x_2} u(x,t) \, dx = \frac{d}{dt} \left[\int_{x_1}^{x_*(t)} u(x,t) \, dx + \int_{x_*(t)}^{x_2(t)} u(x,t) \, dx \right]$$

[3] Typically, everyone uses F for flux instead of G, but we have reserved F for the speed function of propagating interfaces in our level set equation, and it's too late now.

$$= \frac{d}{dt}\left[\int_{x_1}^{x_*(t)} u_1\, dx + \int_{x_*(t)}^{x_2} u_2\, dx\right]$$

$$= \frac{d}{dt}\left[(x_*(t) - x_1)u_1 + (x_2 - x_*(t))u_2\right]$$

$$= u_1\frac{dx_*}{dt} - u_2\frac{dx_*}{dt}$$

$$= (u_1 - u_2)S. \tag{5.21}$$

Thus, using the conservation form (Eqn. 5.20), we have the *Rankine–Hugoniot* condition for the shock speed, given by

$$S = \frac{G(u_2) - G(u_1)}{u_2 - u_1}. \tag{5.22}$$

The plan is to construct weak solutions to the conservation form of Eqn. 5.20 that are viscous limits of the hyperbolic conservation laws. This leads to schemes that will satisfy the conservation form of the equation and will give rise to correct entropy solutions which will remain smooth away from discontinuities.

5.2.2.1 The method of artificial viscosity

One straightforward approach is to approximate numerically the viscous version of the equation. In other words, we solve the equation

$$u_t + [G(u)]_x = \epsilon u_{xx}, \tag{5.23}$$

using a scheme that performs upwinding in the proper direction and relies on the viscosity term to keep things smooth. Thus, recalling our scheme (Eqn. 5.12) that picks out the correct direction of the upwinding, we have the scheme

$$u_i^{n+1} = u_i^n - \Delta t[\max(0, u_i)D^{-x}u_i^n + \min(0, u_i)D^{+x}u_i^n + \epsilon D^{-x}D^{+x}u_i^n]. \tag{5.24}$$

While this scheme works, it is not great. Often, the amount of required artificial diffusion (a large value for ϵ) causes significant rounding at sharp corners. Nonetheless, it has a long history and is used in many settings. As one might guess, many variants exist, in particular those that try to detect and invoke smoothing only where needed.

5.2.2.2 Less diffusive schemes: Lax–Friedrichs

From a practical point of view, a desirable scheme selects the entropy solution and also confines shocks to a few grid points (that is, doesn't smear things out). The previous method of artificial viscosity selects

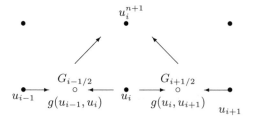

Fig. 5.8. Update of u through numerical flux function g.

the entropy solution, since it adds a diffusive smoothing term. Our goal now is to limit smearing.

Suppose we try to construct schemes that respect the conservation form of the equation. Consider again the conservation law

$$u_t + [G(u)]_x = 0, \tag{5.25}$$

any solution of which also satisfies an integral form of the equation given by

$$\frac{d}{dt} \int_a^b u\,dx = G(u(a,t)) - G(u(b,t)). \tag{5.26}$$

A discrete version of the integral form leads to the following definition:

> **Definition**: A scheme is in *conservation form* if there exists a "numerical flux function" $g(u_{i-1}, u_i)$ $(g(u_i, u_{i+1}))$ which approximates values for $G_{i-1/2}$ $(G_{i+1/2})$ (see Figure 5.8), such that
> $$\frac{u_i^{n+1} - u_i^n}{\Delta t} = -\frac{G_{i+1/2} - G_{i-1/2}}{\Delta x}. \tag{5.27}$$

This definition is natural. It asks that potential scheme at least approximate the hyperbolic conservation law, subject to the consistency requirement $g(u, u) = G(u)$. But how does one also guarantee that the scheme picks out the correct entropy-satisfying weak solution? One answer lies in a further restriction. Consider a scheme W that takes three arguments, the values of $u_i^n, u_{i-1}^n, u_{i+1}^n$ and hands back the value of u at i at the next time step. We say that a 3-point finite difference scheme of the form $u_i^{n+1} = W(u_{i-1}^n, u_i^n, u_{i+1}^n)$ is said to be *monotone* if W is a non-decreasing function of all its arguments. The main fact can now be stated, though we shall not prove it (see Sod [251] and LeVeque [149]):

A conservative, monotone scheme produces a solution that satisfies the entropy condition. Thus, we need only check monotonicity and conservation form in order to verify that a scheme gives the correct entropy condition.

Consequently, this means that to construct a viable scheme, we need only make sure that it is in conservation form and that it is a monotone increasing function of its arguments. One such simple scheme, built from central difference approximations, is called the *Lax–Friedrichs method*, and is given by

$$u_i^{n+1} = \frac{1}{2}[u_{i-1}^n + u_{i+1}^n] - \frac{\lambda}{2}[G_{i+1} - G_{i-1}], \qquad (5.28)$$

where $\lambda = \Delta t/\Delta x$. First, it is straightforward to check that this scheme is monotone if $\frac{dG}{du}\lambda < 1$ and second, it can be put into conservation form by means of the numerical flux function

$$g_{\mathrm{LF}}(u_1, u_2) = -\frac{\Delta x}{2\Delta t}(u_2 - u_1) + \frac{1}{2}[G(u_2) + G(u_1)]. \qquad (5.29)$$

This is a straightforward way of approximating the solution to the general hyperbolic equation. There are other such schemes of this general type, including the Lax–Wendroff method and Fromm's method.

5.2.2.3 Even less diffusive schemes:
exact and approximate Riemann solvers

One advantage of a Lax–Friedrichs scheme is that one need know almost nothing about the structure of the flux function G. One drawback is that it still introduces considerable diffusion into the solution. In other words, sharp discontinuities are smoothed over many number of grid cells, and hence fronts do not stay sharp. In cases where one knows more about the structure of the flux function G, more can be done. In this section, we focus on schemes that keep corners (and hence fronts) as sharp as possible, and limit smearing to only a few grid cells. We will apply them when the flux G is convex[4], that is, when $d^2G/du^2 > 0$.

Given a convex flux function $G(u)$, we want to devise a numerical scheme to solve $u_t + [G(u)]_x = 0$. A fundamental idea, due to Godunov (see [145, 149, 251]), is to take the initial data $\{u_i^n\}$, $i = 1, \ldots, N$, and construct an exact solution to this piecewise constant data at time step $n + 1$ by solving a *local Riemann problem* for each interval. At any time n, stand over each interval at the intermediate grid point $i - 1/2$

---

[4] Things work for non-convex flux functions, with some modification.

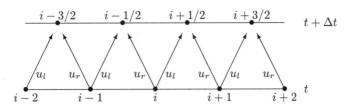

Fig. 5.9. Setup of local Riemann problem.

and consider the local problem given by the solution to the hyperbolic problem with initial data: [5]

$$u(x,0) = \left\{ \begin{array}{ll} u_{\text{left}} = u_{i-1}^n & x < 0 \\ u_{\text{right}} = u_i^n & x \geq 0 \end{array} \right\} \qquad (5.30)$$

The exact solution can be constructed by means of the wave pictures shown previously; we use either a shock, with speed given by the Rankine–Hugoniot condition, or a rarefaction in the case of an expansion wave. The exact solution of each interval's individual Riemann problem is known at time step $n + 1$ at the half grid points; all that remains is to patch the solution together to produce values at the grid points themselves (see Figure 5.9). There are many ways to patch; Godunov constructed a method that averages the solution over staggered intervals and uses that to construct a discrete solution at grid points i at time $n + 1$.

Since the invention of this approach, an array of flux functions g have been developed that solve either exactly or approximately the local Riemann problem to construct the solution at the next time step. The philosophy is always the same. First, make sure that the conservation form of the equation is preserved. Second, make sure that the entropy condition is satisfied. Third, try to give smooth (highly accurate) answers away from the discontinuities. We now follow and refer the reader to an excellent discussion of these issues by Colella and Puckett [69].

One of the easiest such approximate numerical fluxes is the Engquist–

---

[5] Here, the coordinate system for each cell is chosen so for each individual Riemann problem, $x = 0$ corresponds to the point between the two states.

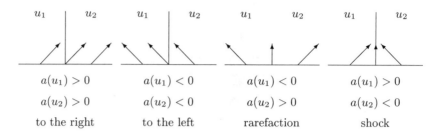

$$
\begin{array}{cccc}
a(u_1) > 0 & a(u_1) < 0 & a(u_1) < 0 & a(u_1) > 0 \\
a(u_2) > 0 & a(u_2) < 0 & a(u_2) > 0 & a(u_2) < 0 \\
\text{to the right} & \text{to the left} & \text{rarefaction} & \text{shock}
\end{array}
$$

Fig. 5.10. Possible solutions to local Riemann problem.

Osher (EO) scheme [83]. This flux is given by

$$
g_{\mathrm{EO}}(u_1, u_2) = G(u_1) + \int_{u_1}^{u_2} \min\left(\frac{dG}{du}, 0\right) du. \tag{5.31}
$$

To see why this is a good scheme, we recall that

$$
u_t + [G(u)]_x = 0, \tag{5.32}
$$

and assume that $G(u)$ is convex; hence $\frac{d^2 G}{du^2} > 0$. Performing the differentiation produces

$$
u_t + \frac{dG}{du} u_x = 0. \tag{5.33}
$$

Thus, $\frac{dG}{du} = a(u)$ is the speed of propagation. Standing in the middle of an interval with a grid point on the left with value u_1 and one on the right with value u_2, we consider all of the cases (see Figure 5.10):

(i) $a(u_1), a(u_2) > 0$: The wave should simply move to the right; indeed, since the integrand is zero, $g_{\mathrm{EO}}(u_1, u_2) = G(u_1)$.

(ii) $a(u_1), a(u_2) < 0$: The wave should simply move to the left; since the integrand is then $\frac{dG}{du}$, integration by the fundamental theorem of calculus produces $g_{\mathrm{EO}}(u_1, u_2) = G(u_2)$.

(iii) $a(u_1) < 0, a(u_2) > 0$: A rarefaction develops, since the speed on the left is negative and those characteristics go to the left, while the speed on the right is positive and hence those characteristics go to the right. The scheme then picks the inverse image of zero; thus $g_{\mathrm{EO}}(u_1, u_2) = G(a^{-1}(0))$.

(iv) $a(u_1) > 0, a(u_2) < 0$: A shock must develop, since the speed on the left is positive and hence those characteristics move to the right, while the speed on the right is negative and hence those characteristics move to the left. The exact solution from the local

Riemann problem depends on the shock speed S; if $S > 0$, then the left value $G(u_1)$ is the answer, while if $S < 0$, then the right value $G(u_2)$ is the answer.

In all except the shock case, this scheme yields the right solution. Let's look more closely at what happens in the shock case (iv). The exact solution is either $G(u_1)$ or $G(u_2)$, depending on the sign of the wave speed S, which comes from the Rankine–Hugoniot condition. However, the scheme gives

$$
\begin{aligned}
g_{\mathrm{EO}}(u_1, u_2) &= G(u_1) + \int_{u_1}^{u_2} \min\left(\frac{dG}{du}, 0\right) du \\
&= G(u_1) + \int_{u_1}^{a^{-1}(0)} \min\left(\frac{dG}{du}, 0\right) du + \int_{a^{-1}(0)}^{u_2} \min\left(\frac{dG}{du}, 0\right) du.
\end{aligned}
$$

The first integral on the right is zero, since the wave speed is positive from u_1 on the left until $a^{-1}(0)$. Thus the solution given by the scheme is

$$
g_{\mathrm{EO}}(u_1, u_2) = G(u_1) + G(u_2) - G(a^{-1}(0)), \tag{5.34}
$$

and see that this scheme is more diffusive than the exact solution. Thus, discontinuities are somewhat smoothed. Fortunately, the characteristics help sharpen things up again, so that the scheme has only a little diffusion. In the specific case of the non-linear wave equation $u_t + [u^2]_x = 0$ one can directly write down the scheme introduced in [187] as

$$
g(u_1, u_2) = (\max(u_1, 0)^2 + \min(u_2, 0)^2). \tag{5.35}
$$

Thus, all the four cases are nicely selected; three of them give the exact solution, and the fourth case adds a little diffusion to the exact solution.

Now that we have a suitable (that is, entropy-satisfying, with relatively little diffusion) difference scheme, we can return to the approximation of the gradients in both the initial value and boundary value formulations. Again, we point out that the subject of schemes to solve hyperbolic conservation laws is vast, and we have ignored many issues involving higher order schemes, convergence issues, and implementation tricks. We refer the interested reader to LeVeque [149], Sod [251], and Colella and Puckett [69] for more comprehensive discussions of the field.

━━━━ 6 ━━━━

Basic Algorithms for Interface Evolution

Outline: *In this chapter, we provide schemes for solving the basic initial and boundary value formulations of interface motion. Our aim is to motivate and present schemes based on the link between Hamilton-Jacobi equations and hyperbolic conservation laws. As we have seen earlier, these two views are formally equivalent for only one-dimensional equations. Nonetheless, the numerical methodology can be used to motivate multi-dimensional schemes for both views of interface propagation. These schemes are followed by a collection of implementation issues.*

6.1 Convergence of schemes for Hamilton-Jacobi equations

We are about to build numerical schemes for Hamilton-Jacobi equations which converge to the viscosity solution. This is, in fact, relatively straightforward to do. Crandall and Lions [74] proved that consistent monotone schemes converge to the correct viscosity solution; this result is analogous to the one for hyperbolic conservation laws. Thus, we could simply write down schemes based on Eqn. 5.35 and then check that they satisfy the necessary requirements for convergence to viscosity solutions. The operators will be easily seen to be consistent, since they use first or higher order finite difference operators and a Taylor series expansion of the error can be shown to go to zero as the time and space step are refined. Proving monotonicity for the most basic first order scheme will be straightforward, simply by checking what happens to monotone data. Higher order schemes are typically not monotone; instead, one uses the notions of monotonicity preserving schemes to show convergence. We refer the interested reader to [74] and [188] for checks of the convergence properties of the higher order schemes presented here and for other schemes as well.

60

However, rather than write schemes down and prove that they satisfy the requirements for convergence, we are instead going to motivate our schemes for Hamilton-Jacobi schemes based on those for hyperbolic conservation laws. This is done for several reasons. First, the reasoning behind upwind methods Hamilton-Jacobi schemes is naturally understood from the perspective of propagating wave direction and characteristics. Second, the straightforward linkage of the one-dimensional equations for front propagation and shocks makes this a natural approach. Third, this upwinding view is critical to the workings of the Fast Marching Method, because it will induce a causality relationship in the ordering of points.

6.2 Hyperbolic schemes and Hamilton-Jacobi equations

6.2.1 One-dimensional schemes

We begin by recalling that both the initial value formulation $\phi_t + F|\nabla \phi| = 0$ and the boundary value formulation $F|\nabla T| = 1$ can often be cast in the form of the general Hamilton-Jacobi equation [1]

$$\alpha U_t + H(U_x, U_y, U_z, x, y, z) = 0. \tag{6.1}$$

The function H is known as the "Hamiltonian"; for the initial value problem we have the Hamiltonian [2]

$$H(U_x, U_y, U_z, x, y, z) = F\sqrt{U_x^2 + U_y^2 + U_z^2}, \tag{6.2}$$

and for the boundary value problem we have the Hamiltonian

$$H(U_x, U_y, U_z, x, y, z) = F\sqrt{U_x^2 + U_y^2 + U_z^2} - 1. \tag{6.3}$$

Let's focus on a one-dimensional version, that is,

$$\alpha U_t + H(U_x) = 0. \tag{6.4}$$

If we let $U_x = u$ and differentiate, we get the hyperbolic conservation law

$$u_t + [H(u)]_x = 0. \tag{6.5}$$

---

[1] Again, we will ignore curvature-driven flows until later in this chapter.
[2] We have trapped ourselves into a notational difficulty. The letter u is typically used for the unknown function in both the Hamilton-Jacobi equation and in hyperbolic conservation laws. Since we are about to go back and forth between the two, we will temporarily let U be the solution of the Hamilton-Jacobi equation, and let u be the solution of the hyperbolic conservation law.

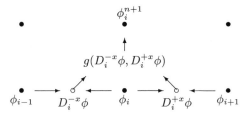

Fig. 6.1. Update of ϕ through numerical Hamiltonian.

In the last chapter, we have seen that numerical fluxes g can be constructed that will accurately approximate this equation through the expression

$$\frac{u_i^{n+1} - u_i^n}{\Delta t} = -\frac{g(u_i^n, u_{i+1}^n) - g(u_{i-1}^n, u_i^n)}{\Delta x}. \tag{6.6}$$

Let's look a little more carefully at this expression in terms of the computational grid shown in Figure 5.8. The value of H at the point $(i-1/2)\Delta x$ (called $H_{i-1/2}$) is approximated by the numerical flux function g as

$$H_{i-1/2} \approx g(u_{i-1}^n, u_i^n). \tag{6.7}$$

Similarly, at the point $i + 1/2$,

$$H_{i+1/2} \approx g(u_i^n, u_{i+1}^n). \tag{6.8}$$

Then from Figure 5.8, the right-hand side of Eqn. 6.6 is just the central difference operator applied to the numerical flux function g. As the grid size goes to zero, consistency requires that $g(u, u) = H(u)$.

This is enough to build a scheme for our one-dimensional Hamilton-Jacobi equation; we can directly write

$$\alpha U_t + H(u) = 0. \tag{6.9}$$

In terms of the computational grid in Figure 6.1, construction of U_i^{n+1} requires U_i^n as well as a value for $H(u_i^n)$. Fortunately, an approximate value for $H(u_i^n)$ is *exactly* what is given by the numerical flux function; thus we have

$$H(u_i^n) \approx g(u_{i-1/2}, u_{i+1/2}). \tag{6.10}$$

All that remains is to construct values for u in the middle of our computational cells. Since $u = U_x$, forward and backward difference approximations in ϕ can be used to construct those values. Thus (see Figure 6.1), we have

$$\alpha U_i^{n+1} = U_i^n - \Delta t\, g\left(\frac{U_i^n - U_{i-1}^n}{\Delta x}, \frac{U_{i+1}^n - U_i^n}{\Delta x}\right), \tag{6.11}$$

where g is one of the numerical flux functions and, again, we have substituted forward and backward difference operators on U for the values of U at the left and right states.

6.2.2 Higher-dimensional schemes

Schemes for higher-dimensional Hamilton-Jacobi equations for symmetric Hamiltonians can be built by simply replicating in each space variable. By building on our one-dimensional case, consider a convex[3] Hamiltonian H. Then using our numerical flux function, we can approximate

$$U_t + H(U_x, U_y, U_z) = 0 \tag{6.12}$$

by

$$\begin{aligned}
U_{i,j,k}^{n+1} = U_{i,j,k}^n - \Delta t\, g\Big(&\frac{U_{ijk}^n - U_{i-1,j,k}^n}{\Delta x}, \frac{U_{i+1,j,k}^n - U_{i,j,k}^n}{\Delta x}, \\
&\frac{U_{ijk}^n - U_{i,j-1,k}^n}{\Delta y}, \frac{U_{i,j+1,k}^n - U_{i,j,k}^n}{\Delta y}, \\
&\frac{U_{ijk}^n - U_{i,j,k-1}^n}{\Delta z}, \frac{U_{i,j,k+1}^n - U_{i,j,k}^n}{\Delta z}\Big).
\end{aligned} \tag{6.13}$$

6.3 The example of a propagating one-dimensional graph

To illustrate these ideas on a simple scheme, let's return to the example of the propagating graph of a function from R to R. In Section 3.3, a central difference approximation was used to track the propagation of a simple corner moving with speed $F = 1$, under the restriction that the

---

[3] In N dimensions, if H is smooth, then H is convex if $\frac{\partial^2 H(u)}{\partial p_i p_j} \geq 0$, where $P = (p_1, \ldots, p_N)$. Alternatively, H is convex if $H(\lambda p + (1-\lambda)q) \leq \lambda H(p) + (1-\lambda)H(q)$ for all $0 \leq \lambda \leq 1$, $p, q \in R^n$. Hamiltonians are not always convex; in Chapter 20, a non-convex speed function which depends on ϕ and ϕ_x, ϕ_y, and ϕ_z will arise naturally out of problems in ion milling and sputter etching.

front remained the graph of a function as it evolved. The equation of motion for the height function ψ was given in Eqn. 2.16 by

$$\psi_t = F(1 + \psi_x^2)^{1/2}. \tag{6.14}$$

The central difference scheme failed because it did not handle the singularity at the corner correctly. Armed with this knowledge, we can write a straightforward scheme. In this case, the Hamiltonian is

$$H(\beta_1) = F(1 + \beta_1^2)^{1/2},$$

and thus we simply apply the scheme given in Chapter 5 with $F = 1$. Figure 6.2 shows the application of the upwind scheme given in Eqn. 6.15. The exact answer is shown, together with two simulations. The first uses the entropy-satisfying scheme with only 20 points (Figure 6.2(b)), the second (Figure 6.2(c)) with 100 points. In the first approximation, the entropy condition is satisfied, but the corner is somewhat smoothed because of the small number of points used. In the more refined calculation, the corner remains sharp, and the exact solution is very closely approximated.

(a) Exact solution   (b) Scheme with 20 Points   (c) Scheme with 100 Points

Fig. 6.2. Upwind, entropy-satisfying approximations to the level set equation.

It is important to point out that far more sophisticated schemes exist than the ones presented here. In the applications of these schemes to hyperbolic problems and shock dynamics, high order resolution schemes are often necessary, because differentiation of the numerical flux function g leads to additional smearing. However, in our case, because we are solving $U_t + H(u) = 0$ rather than $u_t + [H(u)]_x$, the spatial differentiation is not required. For almost all practical purposes, the first and second order schemes presented below are adequate. Complete details of these schemes for advancing the initial and boundary value equations under convex and non-convex Hamiltonians are given below.

6.4 The initial value problem: the Level Set Method

We can now construct the most basic scheme for the initial value problem. We begin with the one-dimensional case, and hence $H(u) = \sqrt{u^2}$. We can once again simply apply the scheme given by Eqn. 5.35 in Chapter 5 and, for speed $F = 1$, write

$$\phi_i^{n+1} = \phi_i^n - \Delta t \, (\max(D_i^{-x}, 0)^2 + \min(D_i^{+x}, 0)^2)^{1/2}. \qquad (6.15)$$

This is the level set scheme given in [187]. (Here, we are using a slightly different shorthand notation; for example, D_i^{-x} means $D_i^{-x}\phi$.) A multi-dimensional version of the level set scheme [187] is then

$$g_{\mathrm{LS}}(u_1, u_2, v_1, v_2, w_1, w_2) = [\,\max(u_1, 0)^2 + \min(u_2, 0)^2 + \qquad (6.16)$$
$$\max(v_1, 0)^2 + \min(v_2, 0)^2 +$$
$$\max(w_1, 0)^2 + \min(w_2, 0)^2]^{1/2}.$$

Thus we have

(i) *First order space convex:*

$$\phi_{ijk}^{n+1} = \phi_{ijk}^n - \Delta t[\max(F_{ijk}, 0)\nabla^+ + \min(F_{ijk}, 0)\nabla^-], \quad (6.17)$$

where

$$\nabla^+ = [\,\max(D_{ijk}^{-x}, 0)^2 + \min(D_{ijk}^{+x}, 0)^2 +$$
$$\max(D_{ijk}^{-y}, 0)^2 + \min(D_{ijk}^{+y}, 0)^2 +$$
$$\max(D_{ijk}^{-z}, 0)^2 + \min(D_{ijk}^{+z}, 0)^2]^{1/2} \qquad (6.18)$$

$$\nabla^- = [\,\max(D_{ijk}^{+x}, 0)^2 + \min(D_{ijk}^{-x}, 0)^2 +$$
$$\max(D_{ijk}^{+y}, 0)^2 + \min(D_{ijk}^{-y}, 0)^2 +$$
$$\max(D_{ijk}^{+z}, 0)^2 + \min(D_{ijk}^{-z}, 0)^2]^{1/2}. \qquad (6.19)$$

Here, we have used a shorthand notation in which $D^{+x}\phi_i^n$ is written as D_i^{+x}, etc.

(ii) *Second order space convex:*

The above schemes can be extended to higher order. The basic trick is to build a switch that turns itself off whenever a shock is detected; otherwise, it will use a higher order approximation to the left and right values by means of a higher order polynomial using an ENO construction (see Harten et al. [110]). These details will not be presented; see [110, 187]. The scheme is the same as before, however, this time ∇^+ and ∇^- are given by

$$
\begin{aligned}
\nabla^+ = [\, \max(A,0)^2 + \min(B,0)^2 + \\
\max(C,0)^2 + \min(D,0)^2 + \\
\max(E,0)^2 + \min(F,0)^2]^{1/2}
\end{aligned}
\tag{6.20}
$$

$$
\begin{aligned}
\nabla^- = [\, \max(B,0)^2 + \min(A,0)^2 + \\
\max(D,0)^2 + \min(C,0)^2 + \\
\max(F,0)^2 + \min(E,0)^2]^{1/2},
\end{aligned}
\tag{6.21}
$$

where

$$
A = D_{ijk}^{-x} + \frac{\Delta x}{2} m(D_{ijk}^{-x-x}, D_{ijk}^{+x-x})
\tag{6.22}
$$

$$
B = D_{ijk}^{+x} - \frac{\Delta x}{2} m(D_{ijk}^{+x+x}, D_{ijk}^{+x-x})
\tag{6.23}
$$

$$
C = D_{ijk}^{-y} + \frac{\Delta y}{2} m(D_{ijk}^{-y-y}, D_{ijk}^{+y-y})
\tag{6.24}
$$

$$
D = D_{ijk}^{+y} - \frac{\Delta y}{2} m(D_{ijk}^{+y+y}, D_{ijk}^{+y-y})
\tag{6.25}
$$

$$
E = D_{ijk}^{-z} + \frac{\Delta z}{2} m(D_{ijk}^{-z-z}, D_{ijk}^{+z-z})
\tag{6.26}
$$

$$
F = D_{ijk}^{+z} - \frac{\Delta z}{2} m(D_{ijk}^{+z+z}, D_{ijk}^{+z-z}),
\tag{6.27}
$$

and the switch function is given by

$$
m(x,y) = \left\{ \begin{array}{l} \left\{ \begin{array}{ll} x & \text{if } |x| \le |y| \\ y & \text{if } |x| > |y| \end{array} \right\} \quad xy \ge 0 \\[2ex] \qquad\qquad 0 \qquad\qquad\quad xy < 0 \end{array} \right\}.
\tag{6.28}
$$

These are the basic level set schemes presented in [187]. We note that these schemes are explicit in time and hence can be programmed in a straightforward manner.

6.4.1 Stability and the CFL condition

There are inherent time step requirements in the above first order time explicit schemes. Analogous with the underlying wave equation, we have a CFL condition which, for an advective speed function F and the first order space scheme, requires the front to cross no more than one grid cell each time step. Thus, we require that

$$\max_{\Omega} F \Delta t \leq \Delta x,$$

where the maximum is taken over values for F at **all** possible grid points, not simply those corresponding to the zero level set.[4] In practice, one can quickly scan the range of values for the speed function F and choose an appropriate time step accordingly.

6.4.2 Higher order time schemes

One way to make these schemes higher order in time is through Runge-Kutta methods. The most straightforward is a second order Heun's method, as follows. Let ϕ_{ijk}^n be the solution at time step $n\Delta t$. We first calculate a temporary value at time $(n+1)\Delta t$ by advancing the level set function one time step using the previous schemes. Thus (using the first order method, for example), let

$$\phi_{ijk}^{(n+1)*} = \phi_{ijk}^n - \Delta t[\max(F_{ijk}^n, 0)\nabla^{n+} + \min(F_{ijk}^n, 0)\nabla^{n-}], \quad (6.29)$$

where we have written F^n and the upwind approximations to the gradients ∇^{n+} and ∇^{n-} with superscripts n to indicate that these are evaluated using the ϕ function at time step $n\Delta t$, and let $\phi_{ijk}^{(n+1)*}$ stand for this temporary update. We then form the actual second order in time update, namely

$$\phi_{ijk}^{(n+1)} = \phi_{ijk}^n - \Delta t/2 \left[\begin{array}{l} \max(F_{ijk}^n, 0)\nabla^{n+} + \min(F_{ijk}^n, 0)\nabla^{n-} + \\ \max(F_{ijk}^{(n+1)*}, 0)\nabla^{*n+} + \min(F_{ijk}^{(n+1)*}, 0)\nabla^{*n-} \end{array} \right],$$
$$(6.30)$$

where the superscript $(n+1)*$ means that we use the temporary value $\phi_{ijk}^{(n+1)*}$ to evaluate F_{ijk}^{n*}, ∇^{n*+}, and ∇^{n*-}. This is a second order method; higher order versions are built in much the same way using higher order Runge-Kutta schemes. Fully implicit versions are somewhat more complex, owing to the subtleties in fully implicit upwind shock schemes.

---

[4] We shall return to this issue later.

6.5 The boundary value problem: the stationary method

We can continue and develop first and second order schemes for the stationary boundary value view. Recall that we wish to solve

$$|\nabla T| F(x, y) = 1.0 \quad \text{such that} \quad \Gamma = \{(x, y)|T(x, y) = 0\}$$

where Γ is the starting position of the interface. Substituting the numerical Hamiltonian from the previous section and Chapter 5, we can immediately produce the following two schemes:

(i) *First order space convex:*

$$\left[\begin{array}{l} \max(D_{ijk}^{-x}T, 0)^2 + \min(D_{ijk}^{+x}T, 0)^2 \\ + \max(D_{ijk}^{-y}T, 0)^2 + \min(D_{ijk}^{+y}T, 0)^2 \\ + \max(D_{ijk}^{-z}T, 0)^2 + \min(D_{ijk}^{+z}T, 0)^2 \end{array} \right]^{1/2} = \frac{1}{F_{ijk}} \quad (6.31)$$

(ii) *Second order space convex:*

$$\left[\begin{array}{l} \max(A, 0)^2 + \min(B, 0)^2 \\ + \max(C, 0)^2 + \min(D, 0)^2 \\ + \max(E, 0)^2 + \min(F, 0)^2 \end{array} \right]^{1/2} = \frac{1}{F_{ijk}} \quad (6.32)$$

where A, B, C, D, E, and F are defined as before.

These schemes are non-linear equations for the grid values T_{ijk}. One approach is to iterate, beginning with an appropriate starting guess. For example, since the first version (Eqn. 6.31) is, in essence, a quadratic equation for the value at each grid point (assuming the others are held fixed), one can iterate until convergence. Typically, one iterates several times through the entire set of grid points until a converged solution is reached (see, for example, [207, 93, 18]). The Fast Marching Method, which will be described in Chapter 8, is a very fast way to solve this set of non-linear equations so that no iteration is required. Instead, an optimal ordering of the grid points is produced so that the equations can be solved with single application of a backsolve.

6.6 Schemes for non-convex speed functions

Given a non-convex speed function F, we can extend our schemes in a variety of ways. One such set of schemes was introduced by Osher and Shu [188]; it results from replacing the Hamiltonian $F|\nabla\phi|$ with the Lax–Friedrichs numerical flux function. For the Level Set Method, this yields the following schemes for the general Hamiltonian $H(u, v, w)$:

(i) *First order space non-convex:*

$$\phi_{ijk}^{n+1} = \phi_{ijk}^n - \Delta t \left[H\left(\frac{D_{ijk}^{-x} + D_{ijk}^{+x}}{2}, \frac{D_{ijk}^{-y} + D_{ijk}^{+y}}{2}, \frac{D_{ijk}^{-z} + D_{ijk}^{+z}}{2} \right) \right. \quad (6.33)$$
$$\left. - \frac{1}{2}\alpha_u(D_{ijk}^{+x} - D_{ijk}^{-x}) - \frac{1}{2}\alpha_v(D_{ijk}^{+y} - D_{ijk}^{-y}) - \frac{1}{2}\alpha_w(D_{ijk}^{+z} - D_{ijk}^{-z}) \right],$$

where α_u (α_v, α_w) is a bound on the partial derivative of the Hamiltonian with respect to the first (second, third) argument, and the non-convex Hamiltonian is a user-defined input function.

(ii) *Second order space non-convex:*

$$\phi_{ijk}^{n+1} = \phi_{ijk}^n - \Delta t[H(\frac{A+B}{2}, \frac{C+D}{2}, \frac{E+F}{2}) \quad (6.34)$$
$$- \frac{1}{2}\alpha_u(B-A) - \frac{1}{2}\alpha_v(D-C) - \frac{1}{2}\alpha_w(F-E)],$$

where A, B, C, D, E, and F are defined as before. For details, see Osher and Shu [188], as well as Adalsteinsson and Sethian [3, 4, 5].

For the stationary boundary value problem, one can pose similar expressions.

6.7 Approximations to geometric variables

As discussed earlier, one advantage of the partial differential equations perspective is that geometric properties of the interface, such as curvature and normal direction, are easily calculated. For example, consider the case of a curve propagating in the plane, formulated as an initial value level set problem. The expression for the curvature of the zero level set assigned to the interface itself (as well as all other level sets) is given by

$$\kappa = \nabla \cdot \frac{\nabla\phi}{|\nabla\phi|} = \frac{\phi_{xx}\phi_y^2 - 2\phi_y\phi_x\phi_{xy} + \phi_{yy}\phi_x^2}{(\phi_x^2 + \phi_y^2)^{3/2}}. \quad (6.35)$$

In the case of a surface propagating in three space dimensions, one has many choices for the curvature of the front, including the mean curvature κ_M and the Gaussian curvature κ_G. Both may be conveniently expressed in terms of the level set function ϕ:

$$\kappa_M = \nabla \cdot \frac{\nabla\phi}{|\nabla\phi|} = \frac{\left\{\begin{array}{c}(\phi_{yy}+\phi_{zz})\phi_x^2 + (\phi_{xx}+\phi_{zz})\phi_y^2 + (\phi_{xx}+\phi_{yy})\phi_z^2 \\ -2\phi_x\phi_y\phi_{xy} - 2\phi_x\phi_z\phi_{xz} - 2\phi_y\phi_z\phi_{yz}\end{array}\right\}}{(\phi_x^2+\phi_y^2+\phi_z^2)^{3/2}},$$

(6.36)

$$\kappa_G = \frac{\left\{\begin{array}{c}\phi_x^2(\phi_{yy}\phi_{zz}-\phi_{yz}^2) + \phi_y^2(\phi_{xx}\phi_{zz}-\phi_{xz}^2) + \phi_z^2(\phi_{xx}\phi_{yy}-\phi_{xy}^2) \\ +2[\phi_x\phi_y(\phi_{xz}\phi_{yz}-\phi_{xy}\phi_{zz}) + \phi_y\phi_z(\phi_{xy}\phi_{xz}-\phi_{yz}\phi_{xx}) \\ +\phi_x\phi_z(\phi_{xy}\phi_{yz}-\phi_{xz}\phi_{yy})\end{array}\right\}}{(\phi_z^2+\phi_y^2+\phi_x^2)^2}.$$

(6.37)

As discussed in [190], this level set formulation can be extended by replacing the Euclidean metric with an arbitrary metric g_{ij}. To do so requires the Christoffel symbols, which can be calculated by

$$\Gamma_{ij}^K = \frac{1}{2}\left(\frac{\partial}{\partial x_i}g_{jl} + \frac{\partial}{\partial x_j}g_{li} - \frac{\partial}{\partial x_l}g_{ij}\right)g^{kl},$$

(6.38)

where g_{ij} is the inverse metric and the usual Einstein summation is assumed. While somewhat time-consuming to evaluate, these expressions extend the level set framework to quite general geometric flows.

Construction of the normal itself can require a more sophisticated scheme than simply building the difference approximation to $\nabla\phi$. This is because the normal can undergo a jump at corners. This suggests the following technique, introduced by Sethian and Strain [243]. First, the one-sided difference approximations to the unit normal in each possible direction are formed. All four limiting normals are then averaged to produce the approximate normal at the corner. Thus, the normal n_{ij} is formed by first letting

$$n_{ij}^* \equiv \frac{\phi_x, \phi_y}{(\phi_x^2 + \phi_y^2)^{1/2}}$$

(6.39)

$$= \frac{(D_{ij}^{+x}, D_{ij}^{+y})}{[(D_{ij}^{+x})^2 + (D_{ij}^{+y})^2]^{1/2}} + \frac{(D_{ij}^{-x}, D_{ij}^{+y})}{[(D_{ij}^{-x})^2 + (D_{ij}^{+y})^2]^{1/2}}$$

$$+ \frac{(D_{ij}^{+x}, D_{ij}^{-y})}{[(D_{ij}^{+x})^2 + (D_{ij}^{-y})^2]^{1/2}} + \frac{(D_{ij}^{-x}, D_{ij}^{-y})}{[(D_{ij}^{-x})^2 + (D_{ij}^{-y})^2]^{1/2}},$$

(6.40)

and then normalizing so that $n_{ij} \equiv n_{ij}^* / |n_{ij}^*|$. If any of the one-sided approximations to $|\nabla \phi|$ is zero over zero, that term is not considered and the weights are adjusted accordingly.

6.8 Calculating additional quantities

The speed function F used to advect an interface can depend on a wide variety of terms. As we have just seen, calculating curvature and normal directions is straightforward. Further geometric properties, such as the length of the interface and the enclosed area, are equally straightforward to calculate, and can in fact expressed in a pure level set framework. One approach laid out and utilized extensively in [282] is as follows. If there is only one interface, then an expression for the enclosed area A is given by

$$A = \int_D H(-\phi(x,y))dxdy, \qquad (6.41)$$

where H is the standard Heaviside function, namely

$$H(x) = \left\{ \begin{array}{ll} 1 & x \geq 0 \\ 0 & x \leq 0 \end{array} \right\}.$$

The length L of the interface is similarly specified, namely

$$L = \int_D \delta(\phi(x,y))|\nabla \phi(x,y)|dxdy \qquad (6.42)$$

where $\delta(x)$ is the delta function. Since both the Heaviside function and the delta function must be numerically approximated, this requires regularized versions which smear their influence over a few grid cells.

When a more accurate evaluation is required, a good approach is to simply find the interface, using any of the standard contour plotters, and then perform an accurate numerical quadrature. Values from the curvature used in the integration along this interface are taken from the grid calculated level set curvature values. If there is more than one interface (in other words, the zero level set of ϕ yields more than one curve), there are two approaches. Again, one can either find the interfaces and then evaluate each separately, or one can modify the Heaviside and delta functions to keep track of which one of the interfaces is involved in the integral. Our experience has been that in applications requiring accuracy, finding the front explicitly, which only requires sweeping through the narrow band (discussed later), is the best approach. Volume-conserving flows requiring accurate evaluation of the enclosed area are discussed and tested in the applications section.

6.9 Initialization

Both the initial value and boundary value approaches require additional data. The level set approach requires an initial function $\phi(x, t = 0)$ with the property that the zero level set of that initial function corresponds to the position of the initial front. A straightforward[5] technique is to compute the signed distance function from each grid point to the initial front. As a practical rule, accuracy is required only near the initial front itself, and a discrete value based on grid distances can suffice far away. In Chapter 11 we discuss how to use Fast Marching Methods to quickly initialize level set problems.

Iterative algorithms for the boundary value formulation require starting guesses which satisfy the boundary condition. If the boundary data (that is, the initial position $\Gamma(t = 0)$ of the front) do not correspond to mesh points, additional care is required.

6.10 Computational domain boundary conditions

The use of a finite computational grid requires boundary conditions. If the speed function F causes the front to expand (such as in the case $F = 1$), upwind schemes will naturally default to outward-flowing one-sided differences at the boundary of the domain, and there is no need for particular attention to boundary conditions. However, in the case of level set problems in which the speed function causes more complex motion, we have usually chosen periodic boundary conditions. These are implemented by creating an extra layer of ghost cells around the domain whose values are simply direct copies of the ϕ values along the actual boundaries. By limiting the given difference schemes to grid points actually on and inside the boundary, the value of ϕ is correctly updated with mirror (reflection) boundary conditions. At the end of each time step, the new values on the boundary are copied to the ghost cells.

---

[5] And expensive.

6.11 Putting it all together

As an example, imagine that we are given an initial closed curve that is evolving under three simultaneous motions. First, it is expanding with a constant speed F_0 in its normal direction. Second, it is moving with speed proportional to its curvature. Third, it is being passively advected by an underlying velocity field $\vec{U}(x, y, t)$ whose direction and strength depend on position and time, but not on the front itself. This entire motion can then be written in terms of the speed function as an explicit level set scheme:

$$F = F_{\text{prop}} + F_{\text{curv}} + F_{\text{adv}}, \tag{6.43}$$

where $F_{\text{prop}} = F_0$ is the propagation expansion speed, $F_{\text{curv}} = -\epsilon\kappa$ is the dependence of the speed on the curvature, and $F_{\text{adv}} = \vec{U}(x, y, t) \cdot \vec{n}$ is the advection speed, where $\vec{n}$ is the normal to the front.

Rather than simply plug this speed function into the schemes, it is a better idea to rearrange terms a bit. Since the normal is given by $\vec{n} = \nabla\phi/|\nabla\phi|$, the level set equation may be re-written as

$$\phi_t + F_0|\nabla\phi| + \vec{U}(x, y, t) \cdot \nabla\phi = \epsilon\kappa|\nabla\phi|. \tag{6.44}$$

The first term on the left (after the time derivative) describes motion in the direction normal to the front and must be approximated through the entropy-satisfying schemes discussed earlier. The second term on the left corresponds to pure passive advection. This term may be approximated through simple upwind schemes. That is, we check the sign of each component of $\vec{U}$ and construct one-sided upwind differences in the appropriate directions. As discussed earlier, the term on the right, which depends on the curvature, is a parabolic contribution to the equation of motion, and hence the use of an upwind scheme, designed for a hyperbolic advection term, is inappropriate. Loosely speaking, this term is like a non-linear heat equation, and information propagates in both directions. Consequently, in terms of our numerical scheme, the most straightforward approach is to use central difference approximations to each of the derivatives in the expression on the right-hand side. For the sake of completeness, we write the complete first order convex scheme to approximate Eqn. 6.44 as

$$\phi_{ij}^{n+1} = \phi_{ij}^n + \Delta t \begin{bmatrix} -[\max(F_{0ij},0)\nabla^+ + \min(F_{0ij},0)\nabla^-] \\ \\ -\left\{ \begin{array}{l} [\max(u_{ij}^n,0)D_{ij}^{-x} + \min(u_{ij}^n,0)D_{ij}^{+x} \\ +\max(v_{ij}^n,0)D_{ij}^{-y} + \min(v_{ij}^n,0)D_{ij}^{+y}] \end{array} \right\} \\ \\ +[\epsilon\, K_{i,j}^n((D_{ij}^{0x})^2 + (D_{ij}^{0y})^2)^{1/2}] \end{bmatrix}$$

$$(6.45)$$

where $\vec{U} = (u,v)$, and $K_{i,j}^n$ is the central difference approximation to the curvature expression given in Eqn. (6.35).

The basic level set and stationary schemes have been presented for both the initial value and boundary value formulations. It is important to point out that, in most cases, far more efficient techniques have been developed. The preferred approaches exploit aspects of adaptivity and result in the Narrow Band Level Set Method for the initial value approach, and Fast Marching Methods for the stationary boundary value view. These are the first two topics addressed in Part III.

Part III

Efficiency, Adaptivity, and Extensions

Part III presents efficient schemes for computing interface motion, based on adaptivity in the case of Narrow Band level set methods, and causality in the case of Fast Marching Methods. After these schemes are discussed, both methodologies are moved to triangulated unstructured grids. Algorithms to couple these techniques to applications are provided, followed by basic numerical tests of the methods, and then a few remarks about implementation.

7

Efficient Schemes: the Narrow Band Level Set Method

Outline: *The Level Set Method presented above is a relatively straightforward version that may be easily programmed. However it is not particularly fast, nor does it make efficient use of data structures and computational resources. In this chapter, we consider more sophisticated versions of the basic scheme.*

7.1 Parallel algorithms

The straightforward approach is to solve the initial value partial differential equation for the level set function ϕ in the entire computational domain. We call this a "full matrix approach," since one is updating *all* the level sets, not just the zero level set corresponding to the front itself. The advantage of this approach is that the data structures and operations are extremely clear, and it is a good starting point for building level set codes.

There are circumstances in which this approach is desirable. If all the level sets are themselves important (such as in problems encountered in image processing discussed in a later section), then computation over the entire domain is required. In this case, one simple speedup is obtained through a parallel computation. Since each grid point is updated by a nearest neighbor stencil using only grid points on each side, this technique is close to "embarrassingly parallel." A parallel version of the Level Set Method was developed by Sethian [227] for the Connection Machine CM-2 and CM-5. In the CM-2, nodes are arranged in a hypercube fashion; in the CM-5, nodes are arranged in a fat-tree. The code was written in global CMFortran, and at each grid point CSHIFT operators were used to update the level set function. A time-explicit second

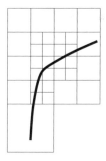

Fig. 7.1. Cell hierarchy.

order space method was used to update the level set equation. Output was controlled by linking the level set evolution to a parallel volume rendering routine, with associated display through access to a parallel frame buffer. As expected, the operation count per time step reduces to $O(1)$, since in most cases the full grid can be placed into physical memory. Thus, most applications of updating propagating interfaces according to given speed functions transpire as real-time movies, the main limitation being the speed of display.

7.2 Adaptive mesh refinement

One version of an efficient Level Set Method comes from pursuing an adaptive mesh refinement strategy. This is the approach taken by Milne [178], motivated by the adaptive mesh refinement work by Berger and Colella [26]. Adaptivity may be desired in regions where level curves develop high curvature or where speed functions change rapidly; if the zero level curve identified with a front is the object of interest, then the mesh can be adaptively refined around its location. To illustrate this approach, Figure 7.1 shows mesh cells that are hierarchically refined around a large curvature in the zero level set of ϕ. Calculations are performed on both the fine grids and the coarse grids. The grid cell boundaries always lie along $x - y - z$ coordinate lines and patches do not overlap; no attempt is made to align the refined cells with the front.

The data structures for the adaptive mesh refinement are fairly straightforward. However, considerable care must be taken at the interfaces between coarse and fine cells. In particular, a subtle update strategy for ϕ is required at so-called "hanging nodes" where the boundary between two levels of refinement do not have a full set of nearest neigh-

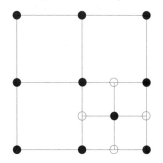

Fig. 7.2. Grid values at boundary between refinement levels.

bors. To illustrate, Figure 7.2 shows a two-dimensional adaptive mesh; the goal is to determine an accurate update strategy for the hanging nodes marked "o".

The strategy laid out by Milne for updating ϕ at such points is as follows. Consider the archetypical speed function $F(\kappa) = 1 - \epsilon\kappa$.

- The advection term 1 leads to a hyperbolic equation. Straightforward interpolation of the updated values of ϕ from the coarse cell grid is used to produce the new value of ϕ at o. More sophisticated technology is not required, since we are modeling the update according to the numerical flux function g, not the derivative of the numerical flux function as required for hyperbolic conservation laws.

- In the case of the curvature term $-\epsilon\kappa$, the situation is not so straightforward, since this corresponds to a parabolic term that cannot be approximated through simple interpolation. Milne showed that straightforward interpolation from updated values on the coarse grid to the fine grid provides poor answers. Using this procedure, the boundary between the two levels of refinement acts as a source of noise, and significant error is generated at the boundary. When tested on the simple heat equation using a coarse/fine mesh, this approach produces more error in the computed solution than would be produced using a coarse mesh everywhere. Instead, Milne devised the following technique. Values from both the coarse grid and refined grid around the hanging node are used to construct a least squares solution for ϕ before the update. This solution surface is then formally differentiated to produce the various first and second derivatives in each component direction. These values are then used to produce the update value for ϕ in the same manner as all other nodes.

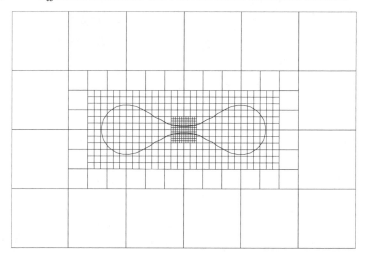

Fig. 7.3. Two-dimensional slice of adaptive mesh for propagating surface.

As illustration, Figure 7.3 shows a two-dimensional slice of a fully three-dimensional adaptive mesh calculation of a surface collapsing under its mean curvature. As will be discussed in a later section, because one principal axis of curvature is very large and positive, the dumbbell neck pinches off and breaks into two.

7.3 Narrow banding and fast methods

When one is interested in only a specific front, there are several disadvantages with the full matrix approach. As formulated, each grid point contains the value of the level set function at that point, and thus there is an entire family of contours, only one of which is the zero level set (see Figure 7.4). The level set method stands at each grid point and updates its value to correspond to the motion of the surface. This produces a new contour value at that grid point.

A more efficient approach is to work only near the front of interest. There are several reasons to do so.

- *Speed:* Performing calculations over the entire computational domain requires $O(N^2)$ operations per time step in two dimensions, and $O(N^3)$ operations in three dimensions, where N is the number of grid points along a side. As an alternative, an efficient modification is to perform work only in a neighborhood of the zero level set; this is known as

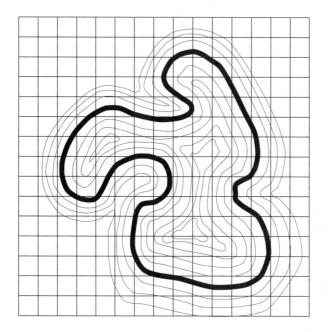

Fig. 7.4. Dark line is zero level set corresponding to front.

the *Narrow Band approach*. If we assume that the front has roughly $O(N^2)$ points in three dimensions, then the operation count in three dimensions drops to $O(kN^2)$, where k is the number of cells in the narrow band, a significant cost reduction.

- *Calculating Extension Variables*: As will be discussed in detail in Chapter 11, the level set approach requires a speed function F defined on *all* of the domain; not simply on the zero level set corresponding to the front. Recall that three types of arguments may influence the front speed F: local, global, and independent. Some of these variables may have meaning only on the front itself, and it may be both difficult and awkward to design a speed function that extrapolates the velocity away from the zero level set in a smooth fashion. Thus, another advantage of the Narrow Band approach is that this extension need be done only for points lying near the front, as opposed to all points in the computational domain.

- *Time Steps*: The full-matrix approach requires a time step that sat-

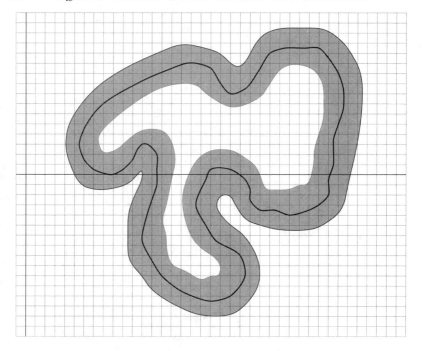

Fig. 7.5. Grid points in dark area are members of narrow band.

isfies a CFL condition with regard to the maximum velocity over the entire domain, not simply in response to the speed of the front itself. In a narrow band implementation, the time step can be adaptively chosen in response to the maximum velocity field achieved within the narrow band. This is advantageous when the front speed changes substantially as it moves (such as in curvature flow). In such problems, the CFL restriction for the velocity field for *all* the level sets may be much more stringent than the one for those sets within the narrow band.

The above "Narrow Band Method" was introduced by Chopp [59], used in recovering shapes from images by Malladi, Sethian and Vemuri [168], and analyzed extensively by Adalsteinsson and Sethian [2]. The idea is straightforward and can be understood by means of two figures.

Figure 7.5 shows the placement of a narrow band around the familiar initial front. The entire two-dimensional grid of data is stored in a square array. A one-dimensional array is then used to keep track of the points in this narrow band (shaded grid cells in Figure 7.5 are located in a narrow

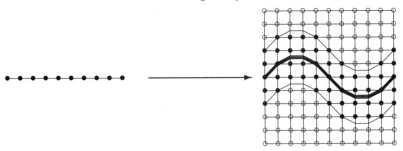

Fig. 7.6. Pointer array tags interior and boundary band points.

band around the front of a user-defined width); see Figure 7.6. Only the values of ϕ within the narrow band are updated. Values of ϕ at grid points on the boundary of the narrow band are frozen. When the front moves near the edge of the tube boundary, the calculation is stopped, and a new tube is built with the zero level set interface boundary at the center. This rebuilding process is known as "re-initialization." Because the front does not move all the way to the boundary, these boundary conditions do not adversely affect the motion of the zero level set.

Thus, the Narrow Band Method consists of the following loop:

(i) Tag *Alive* points in narrow band.
(ii) Build *Land Mines* to indicate near edge.
(iii) Initialize *Far Away* points outside the narrow band with large positive (negative) values if values are outside (inside) the front itself.
(iv) Solve level set equation until *Land Mine* hit.
(v) Rebuild and loop.

Use of narrow bands leads to level set front advancement algorithms that are equivalent in terms of complexity to traditional marker methods and cell techniques, while maintaining the advantages of topological merger, accuracy, and easy extension to multiple dimensions. Typically, the Narrow Band Level Set Method is about ten times faster on a 160×160 grid than the full matrix method. Such a speedup is substantial, and in three-dimensional simulations can make the difference between computationally intensive problems and those that can be done with relative ease.

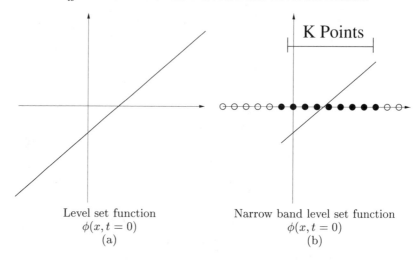

Level set function
$\phi(x, t = 0)$
(a)

Narrow band level set function
$\phi(x, t = 0)$
(b)

Fig. 7.7. Full level set and narrow band formulations.

7.4 Details of the Narrow Band implementation

We now discuss the Narrow Band approach in a little more detail. Consider the simple problem of a point, initially at $x = 1$, moving with speed F along the positive x-axis. The "Langrangian formulation" of this problem requires solving the ordinary differential equation

$$\frac{dx(t)}{dt} = F \qquad x(t = 0) = 1.$$

Alternatively, the level set formulation, initialized with the signed distance function, requires solving the partial differential equation[1]

$$\phi_t(x, t) + F|\phi_x| = 0 \qquad \phi(x, t = 0) = x - 1.$$

Rather than use all the grid points, the Narrow Band approach stores values for ϕ at grid points only within the narrow band. Figure 7.7a shows the full level set function, while Figure 7.7b shows a level set function defined only on a narrow band of width K points. We update the grid values in this narrow band. When the zero level set reaches the edge of the narrow band, we stop and rebuild.

As discussed previously, the data structure for this Narrow Band implementation is very straightforward. Making the Narrow Band Method of higher order is also straightforward, since the update is unconcerned

---

[1] Of course, it is silly to frame this problem as a level set method; we set it up only in order to explain further the Narrow Band approach.

with the band structure. A more subtle issue is the choice of an appropriate width k for the narrow band. We consider the two extremes. If the narrow band is infinitely large, it is never rebuilt, and this technique becomes equivalent to the full matrix approach. No computational speedup is obtained. At the other extreme, if the narrow band is three grid points across, one must re-initialize (that is, build a new signed distance function from the zero level set) almost every time step, and the cost of this re-initialization becomes significant. Our experience indicates that a narrow band width of about six grid points on either side of the zero level set is a reasonable balance between re-initialization costs and update costs. Further details on the accuracy, typical tube sizes, and number of times a tube must be rebuilt may be found in Adalsteinsson and Sethian [2]. Details about techniques for re-initialization and associated error are described in Chapter 11.

There are several variants of our Narrow Band Level Set Method. For example, rather than pick a fixed size for the narrow band, one might decide to pick a thin band where the speed function F is small and increase its size at places where the speed function F is large. This will have the effect that the interface moves towards the edge of the narrow band more uniformly.

Another variant is to characterize the narrow band by the values of the level set function ϕ, rather than by distance in the domain space, and dynamically add grid points to the narrow band as it moves based on the values of the level set function. Points whose $|\phi|$ values go above a certain threshold level are removed, while neighbors are added around those that dip below the threshold. When new grid points are added, they must be given appropriate ϕ values. This is accomplished by re-initializing every time step to return to the signed distance function. As new grid points are added, the ϕ function in the entire narrow band is re-initialized, and the calculation is advanced one time step. This variant on our Narrow Band method was used in [54, 262] and requires constant re-initialization. Further discussion of the effects of re-initialization and the accuracy of Narrow Band methods will be discussed in Chapters 11 and 12.

8

Efficient Schemes: Fast Marching Methods

Outline: *The standard methods for the boundary value view require iteration. We now describe Fast Marching Methods, which allow one to solve the boundary value problem without iteration. To do so, an optimal ordering of the grid points is produced so that one can solve the non-linear system with a single application of a backsolve.*

Our goal is to construct algorithms to solve the Eikonal equation[1] $|\nabla T|F = 1$. Recall our one-sided difference scheme given in Eqn. 6.31, namely,

$$
\left[
\begin{array}{l}
\max(D_{ijk}^{-x}T, 0)^2 + \min(D_{ijk}^{+x}T, 0)^2 \\
+ \max(D_{ijk}^{-y}T, 0)^2 + \min(D_{ijk}^{+y}T, 0)^2 \\
+ \max(D_{ijk}^{-z}T, 0)^2 + \min(D_{ijk}^{+z}T, 0)^2
\end{array}
\right]^{1/2}
= \frac{1}{F_{ijk}}. \tag{8.1}
$$

A slightly different upwind scheme, given in [207], will turn out to be more convenient, namely,

$$
\left[
\begin{array}{l}
\max(D_{ijk}^{-x}T, -D_{ijk}^{+x}T, 0)^2 \\
+ \max(D_{ijk}^{-y}T, -D_{ijk}^{+y}T, 0)^2 \\
+ \max(D_{ijk}^{-z}T, -D_{ijk}^{+z}T, 0)^2
\end{array}
\right]^{1/2}
= \frac{1}{F_{ijk}}. \tag{8.2}
$$

---

[1] Before we start, we point out a slightly confusing issue which in fact is at the heart of the efficiency of Fast Marching Methods. The Eikonal equation is a boundary value problem. However, we are going to devise numerical schemes that will efficiently build the solution outward from the boundary data. While this sounds like an initial value problem, it is only our algorithmic construction that has this propagation view.

8.1 Iteration

How might one solve Eqn. 8.2? One solution, as given in [207], is through iteration. Consider a stencil of a grid point and its six neighbors, as shown in Figure 8.1.

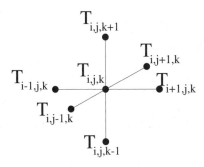

Fig. 8.1. Updating a grid point.

Observe that Eqn. 8.2 is a quadratic equation for T_{ijk}, assuming that the neighboring grid values for T are given. Thus, one solution comes from updating the value of u at each grid point according to this quadratic until a solution is reached:

> For iter=1,n;
>> For i,j,k=1,dim
>>> Solve Quadratic for $T_{ijk}^{\text{iter}+1}$, given
>>>
>>> $$T_{i-1,j,k}^{\text{iter}}, T_{i+1,j,k}^{\text{iter}}, T_{i,j-1,k}^{\text{iter}}, T_{i,j+1,k}^{\text{iter}}, T_{i,j,k-1}^{\text{iter}}, T_{i,j,k+1}^{\text{iter}} \qquad (8.3)$$
>> EndFor
> EndFor

An operation count on this approach, assuming N points in each direction, yields at least $O(N^4)$ labor. Here, we are taking an optimistic guess of roughly N steps to converge; in reality, iteration can often take much longer to converge.

The key to Fast Marching Methods lies in the observation that this iteration contains a very specific causality relationship. We now describe the Fast Marching Method; for details, see [233, 235].

8.2 Causality

The central idea behind the Fast Marching Method is to systematically construct the solution T using only upwind values. We observe that the

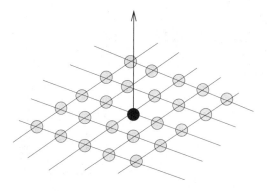

Fig. 8.2. Beginning of Fast Marching Method.

upwind difference structure of Eqn. 8.2, (as well as Eqn. 8.1), allows us to propagate information "one way," that is, from smaller values of T to larger values. The Fast Marching Method rests on building the solution of Eqn. 8.2 outward from the smallest T value, stepping away from the boundary condition in a downwind direction. We sweep the front along by considering points in a thin zone around the existing front, and marching this thin zone forward, freezing the values of existing points and bringing new ones into the narrow band structure. The key is in the selection of *which* grid point in the narrow band to update.

Consider a two-dimensional version of the Eikonal equation $|\nabla T|F = 1$ in which the boundary value is known at the origin, shown as a black sphere in Figure 8.2. The light gray spheres are grid points where the solution value is unknown.

We start the algorithm by solving Eqn. 8.2 by marching "downwind" from the known value, computing new values at each of the four neighboring grid points; as shown in Figure 8.3. This provides possible values for u at each grid point $u_{-1,0}, u_{1,0}, u_{0,-1}, u_{0,1}$, shown as dark gray spheres in Figure 8.3.

We now would like to march downwind from these values given at the dark gray spheres. *Observe that dark gray sphere with the smallest T must have the correct value.* This is because upwinding dictates that no point can be affected by grid points containing larger values of T. Thus, we may freeze the value of T at this smallest dark gray sphere[2] and proceed with the algorithm, as shown in Figure 8.3.

Since recomputing the T values at downwind neighboring points can-

---

[2] that is, turn it into a black sphere and consider its value known.

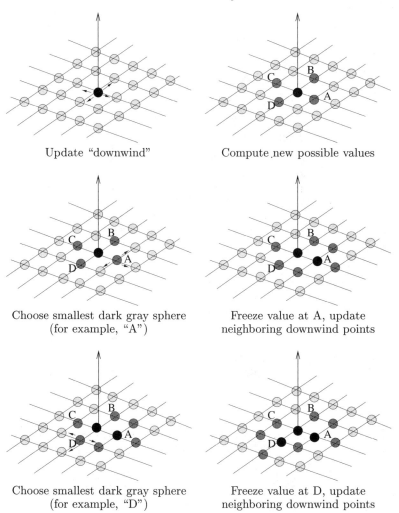

Update "downwind"   Compute new possible values

Choose smallest dark gray sphere   Freeze value at A, update
(for example, "A")   neighboring downwind points

Choose smallest dark gray sphere   Freeze value at D, update
(for example, "D")   neighboring downwind points

Fig. 8.3. Update procedure for Fast Marching Method.

not yield a value smaller than that at any of the known points, we can systematically move forward. We march the solution outwards, always transforming the dark gray grid point with minimum trial value for T into a known value, and readjusting downwind neighbors (see Figure 8.4). This means that we need never go back and "revisit" a point with a known value; that value will remain unchanged by all later calculations. Another way to look at our algorithm is that each minimum trial value

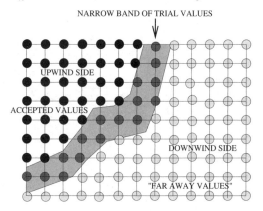

Fig. 8.4. Progress of Fast Marching Method.

begins an application of Huygens' principle, and the expanding wave front touches and updates all others. For more details, see [235, 233].

8.3 The update procedure for the Fast Marching Method

Thus, the Fast Marching Method is as follows: First, tag points the boundary value points as *Known*. Then tag as *Trial* all points that are one grid point away. Finally, tag as *Far* all other grid points. Then the loop is as follows:

   (i) Begin loop: Let A be the *Trial* point with the smallest T value.
   (ii) Add the point A to *Known*; remove it from *Trial*.
   (iii) Tag as *Trial* all neighbors of A that are not *Known*. If the neighbor is in *Far*, remove, and add to the set *Trial*.
   (iv) Recompute the values of T at all *Trial* neighbors of A according to Eqn. 8.2 by solving the quadratic equation.
   (v) Return to top of loop.

8.4 Heap sorts and computational efficiency

The key to an efficient version of this technique lies in devising a fast way of locating the grid point in the narrow band with the smallest value for T. An efficient scheme to do so is discussed in [233]; here we follow that discussion.[3]

---

[3] The work and contributions of Dr. Ravikanth Malladi were invaluable in building the initial version of the Fast Marching Method.

We use a min–heap data structure (see Sedgewick [221]) with back pointers to store the T values. In an abstract sense, a min-heap is a "complete binary tree" with the property that the value at any given node is less than or equal to the values at its children. In practice, it is more efficient to represent a heap sequentially as an array by storing a node at location k and its children at locations $2k$ and $2k + 1$. From this definition, the parent of a given node at k is located at $k/2$. Therefore, the root which contains the smallest element is stored at location $k = 1$ in the array. Finding the parent or children of a given element are simple array accesses which take $O(1)$ time.

The values of T are stored, together with the indices which give their locations in the grid structure. The marching algorithm works by first looking for the smallest element in the *NarrowBand*; this FindSmallest operation involves deleting the root and using one sweep of DownHeap to ensure that the remaining elements satisfy the heap property. The algorithm proceeds by tagging the neighboring points that are not *Alive*. The *FarAway* neighbors are added to the heap using an Insert operation, and values at the remaining points are updated using Eqn. 8.2. Insert works by increasing the heap size by one and trickling the new element upward to its correct location using an UpHeap operation. Lastly, to ensure that the updated u values do not violate the heap property, we need to perform an UpHeap operation starting at that location and proceeding up the tree.

The DownHeap and UpHeap operations (in the worst case) carry an element all the way from root to bottom or vice versa. Therefore, this takes $O(\log M)$ time assuming there are M elements in the heap. It is important to note that the heap, which is a complete binary tree, is always guaranteed to remain balanced. All that remains is the operation of searching for the *NarrowBand* neighbors of the smallest element in the heap. This can be made $O(1)$ in time by maintaining back pointers from the grid to the heap array. Without the back pointers, the search takes time $O(M)$ in the worst case.

As an example, Figure 8.5 shows a typical heap structure and an UpHeap operation after the element at location $(2, 7)$ gets updated from 3.1 to 2.0.

Thus, since the total work in changing the value of one element of the heap and bubbling its value upward is $O(\log M)$, where M is the size of the heap, this produces a worst case total operation count of $M \log M$ for the Fast Marching Method on a grid of M total points. If we imagine a three-dimensional grid of N points in each direction, the Fast Marching

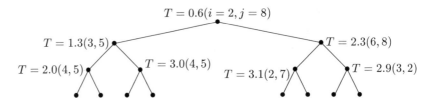

Step 1: Change u value at (2,7)

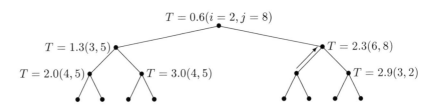

Step 2: New value at (2,7); UpHeap

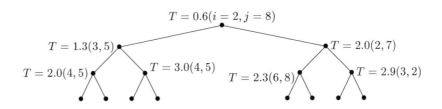

Heap property restored

Fig. 8.5. Heap structure and UpHeap+ operation.

Method reduces the total operation count to $N^3 \log N$. Essentially, each grid point is visited once to compute its arrival time value.

8.5 Initial conditions

The above technique considered a flat initial interface for which trial values at the narrow band points could be easily initialized. Given an arbitrary closed curve or surface as the initial location of the front, one can use the original Narrow Band Level Set Method to initialize the problem. First, label all grid points as *Far Away*, and assign them T

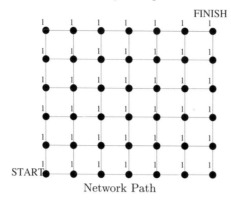

Fig. 8.6. Network path problem.

values of ∞. Then, in a very small neighborhood around the interface, construct the signed distance function from the initial hypersurface Γ. Propagate that surface both forward and backward in time under the speed law F until a layer of grid points is crossed in each direction, computing the signed crossing times as in [230]. Then collect the points with negative crossing times as *Known* points with T value equal to the crossing times, and the points with positive crossing times as *Trial* points with T value equal to the positive crossing times. Then begin the Fast Marching Method. A more sophisticated technique for initializing the Fast Marching Method is given in [6].

8.6 Network path algorithms

The Fast Marching Method is reminiscent of Dijkstra's algorithm [78], which is a method for finding the shortest path on a network with prescribed weights for each link between nodes. As illustration, imagine one is given a rectangular network with an equal unit cost for each link (see Figure 8.6).

Dijkstra's method takes a starting point and expands outward, visiting each node and storing a sum of the total cost. The "front" is advanced by looking for the node reached with the smallest current cost, and then advancing to neighbors. In this sense, the method is similar to the Fast Marching Method. Suitably programmed by means of a heap as described above, the method is an $O(N \log N)$ method of computing the cost of going from point A to any other point in a network of n nodes.

However, if the two points are positioned relative to the network so

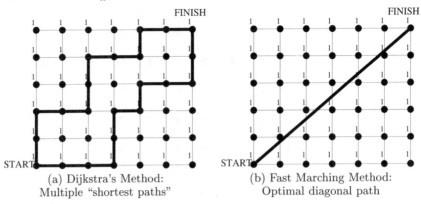

(a) Dijkstra's Method: (b) Fast Marching Method:
Multiple "shortest paths"   Optimal diagonal path

Fig. 8.7. Inconsistent vs. consistent algorithms.

that the optimal path is a straight line diagonal between the two, such a graph search algorithm cannot distinguish among the various network graphs of equal cost connecting points A and B. In our front evolution view, the actual problem one wants to approximate is the solution of the **continuous** problem, not the network solution. Dijktra's method is inconsistent with the underlying continuous problem. Refinement of the network will not produce a solution which converges to the correct diagonal shortest path. While it is possible to add diagonal links to the network to produce the diagonal using Dijkstra's method, the Fast Marching Method provides an approach which directly approximates the solution of the underlying partial differential equation through consistent numerical approximations. It finds the correct diagonal with relatively few nodes and with no additional computational labor.

8.6.1 The Link between Dijkstra's method and Huygens' principle

We can dig a little deeper and reframe Dijkstra's method in the terminology of Huygens' principle. Suppose we consider a rectangular mesh with a "cost function" $\frac{1}{F_{ij}}$ defined at each mesh point. Here, we consider a particle moving along the edges connecting these grid points, and the cost function is the amount of time the particle must wait at a node before being allowed to move on. If F_{ij} is large, then the particle does not need to wait long; as F decreases, the particle must wait longer. Of all possible paths from start to finish, our goal is to find the one that requires the least amount of total time.

Dijkstra's method fans out from the starting point in all possible paths, computing a running temporary total cost for each new grid point as it is reached. The update rule to evaluate this temporary trial cost T^{trial} at the mesh point (i, j) is

$$T_{ij}^{\text{trial}} = \min\left(T_{i-1,j}, T_{i+1,j}, T_{i,j-1}, T_{i,j+1}\right) + \frac{1}{F_{ij}}. \qquad (8.4)$$

In other words, we look at the total time required to reach each neighbor, and then take the smallest one, plus the time required to reach the node itself, as the new total time T_{ij}^{temp} at the grid point (i, j). Dijkstra's method, similar to the Fast Marching Method, then chooses the smallest of all such trial values, converts it to "known", and then fans out to produce trial updates T^{trial} at all the neighbors of the known point, according to Eqn. 8.4.

This "fanning out" from the smallest temporary point is the equivalent of Huygens' principle. We imagine a wave front expanding from each point, and the process of taking the envelope of that wave front is equivalent to the use of the minimum value in Eqn. 8.4. In the case of Dijkstra's method, the fanning out is confined to the network path edges; in the case of the continuous problem solved by the Fast Marching Method, the fanning out occurs with a sub-grid approximation to the continuous problem. We can make the link more explicit by rewriting Eqn. 8.4 with forward and backward difference schemes as follows:

With surrounding values fixed, let T_{ij}^{trial} be the smallest value of T_{ij} such that

$$\max(D_{ijk}^{-x}T, -D_{ijk}^{+x}T, D_{ijk}^{-y}T, -D_{ijk}^{+y}T, 0) = \frac{1}{F_{ijk}}. \qquad (8.5)$$

Thus, we can see what Dijkstra's method is doing. Given a rectangular mesh, Dijkstra's method produces a non-unique solution to the equation

$$\max|u_x, u_y| = \frac{1}{F_{ijk}},$$

while the Fast Marching Method finds the unique viscosity solution of the static Hamilton-Jacobi equation

$$\left(u_x^2 + u_y^2\right)^{1/2} = \frac{1}{F_{ijk}}.$$

Both rely on a Huygens' principle wave front and use the minimum value update idea, however the Fast Marching Method is consistent with the underlying viscosity solution of the partial differential equation.

8.7 Optimal orderings

A different way to look at the Fast Marching Method is by returning to the iterative loop given in Eqn. 8.3, namely

> For iter=1,n;
>
>> For i,j,k=1,dim
>>
>>> Solve Quadratic for $T_{ijk}^{\text{iter}+1}$, given
>>>
>>> $$T_{i-1,j,k}^{\text{iter}}, T_{i+1,j,k}^{\text{iter}}, T_{i,j-1,k}^{\text{iter}}, T_{i,j+1,k}^{\text{iter}}, T_{i,j,k-1}^{\text{iter}}, T_{i,j,k+1}^{\text{iter}} \qquad (8.6)$$
>>
>> EndFor
>
> EndFor.

The Fast Marching Method is a re-ordering of the points so that the inner loop becomes a straightforward backsolve, and hence no outer loop iteration is necessary. The price of computing the correct re-ordering, done while the loop is in process, is the $O(\log N)$ steps required to re-order the heap as the values as changed.

8.8 Higher accuracy Fast Marching Methods

As presented, the Fast Marching Method is a first order scheme, owing to the use of a first order approximation to the gradient, namely,

$$\left[\begin{array}{l} \max(D_{ijk}^{-x}T, -D_{ijk}^{+x}T, 0)^2 + \\ \max(D_{ijk}^{-y}T, -D_{ijk}^{+y}T, 0)^2 + \\ \max(D_{ijk}^{-z}T, -D_{ijk}^{+z}T, 0)^2 \end{array} \right]^{1/2} = \frac{1}{F_{ijk}}. \qquad (8.7)$$

We now produce a way of designing Fast Marching Methods with higher accuracy.

We begin by noting a slightly different implementation of the Fast Marching Method:[4] when a *Trial* trial value (which are the tentative values in the heap) is being recomputed, use only *Known* values in the computation. We now introduce a way of using second order operators in the Fast Marching Method. We first note that a second order backward approximation to the first derivative T_x is given by

$$T_x \approx \frac{3T_i - 4T_{i-1} + T_{i-2}}{2\Delta x},$$

which may be compactly written as

$$T_x \approx D^{-x}T + \frac{\Delta x}{2} D^{-x-x}T.$$

---

[4] This different coding was suggested by D. Adalsteinsson.

A similar expression holds for the forward difference, namely

$$T_x \approx D^{+x}T - \frac{\Delta x}{2}D^{+x+x}T.$$

Consider now the switch functions defined by (the expressions are similar in y and z)

$$\text{switch}_{ijk}^{-x} = \begin{bmatrix} 1 & \text{if } T_{i-2,j,k} \text{ and } T_{i-1,j,k} \text{ are known and } T_{i-2,j,k} \le T_{i-1,j,k} \\ 0 & \text{otherwise} \end{bmatrix},$$

$$\text{switch}_{ijk}^{+x} = \begin{bmatrix} 1 & \text{if } T_{i+2,j,k} \text{ and } T_{i+1,j,k} \text{ are known and } T_{i+2,j,k} \le T_{i+1,j,k} \\ 0 & \text{otherwise} \end{bmatrix}.$$

We can then use these operators in the Fast Marching Method, namely,

$$\left[\begin{array}{l} \max \left[\left[D_{ijk}^{-x}T + \text{switch}_{ijk}^{-x}\frac{\Delta x}{2}D_{ijk}^{-x-x}T \right], -\left[D_{ijk}^{+x}T - \text{switch}_{ijk}^{+x}\frac{\Delta x}{2}D_{ijk}^{+x+x}T \right], 0 \right]^2 \\ + \\ \max \left[\left[D_{ijk}^{-y}T + \text{switch}_{ijk}^{-y}\frac{\Delta y}{2}D_{ijk}^{-y-y}T \right], -\left[D_{ijk}^{+y}T - \text{switch}_{ijk}^{+y}\frac{\Delta y}{2}D_{ijk}^{+y+y}T \right], 0 \right]^2 \\ + \\ \max \left[\left[D_{ijk}^{-z}T + \text{switch}_{ijk}^{-z}\frac{\Delta z}{2}D_{ijk}^{-z-z}T \right], -\left[D_{ijk}^{+z}T - \text{switch}_{ijk}^{+z}\frac{\Delta z}{2}D_{ijk}^{+z+z}T \right], 0 \right]^2 \end{array} \right]^{1/2}$$

$$= \frac{1}{F_{ijk}}.$$

$$(8.8)$$

This scheme attempts to use a second order one-sided upwind stencil whenever points are available, but reverts to a first order scheme in the other cases. We note two points. First, in order to start the scheme, we must use a higher order scheme to produce accurate values in a band around the boundary values. Second, we have chosen a very "conservative" higher order version of our Fast Marching Method; it is possible to devise versions which invoke the first order scheme less often.

8.8.1 Accuracy/order of this scheme

At this point, it is reasonable to ask if this really is a second order scheme. If a scheme reverts to first order at some point, doesn't that render the entire error first order? The answer is, in fact, "not necessarily", for the situation is more subtle. As an example, consider an integration scheme to solve the simple ordinary differential equation

$$\frac{dy}{dx} = f(x, y) \qquad y(x = 0) = \text{given.}$$

Suppose we want the value of $y(x)$ at $x = 10$, and suppose further that we use a second order scheme to integrate from $x = 0$ to $x = 5$, then a first order scheme for one space step Δx from $x = 5$ to $x = 5 + \Delta x$, and then a second order scheme the rest of the way to $x = 10$. This in fact is a global second order scheme. To see that this is so, we observe that the scheme error from $x = 0$ to $x = 5$ is locally third order, and globally second order, as is the scheme from $x = 5 + \Delta x$ to $x = 10$. The error from $x = 5$ to $x = 5 + \Delta x$ is locally second order. Hence the total error is globally second order. The point in this (admittedly rigged) example is that, as the mesh is refined, the total number of points where the first order scheme is invoked does not increase.

Thus, the question of whether or not the "second order" Fast Marching Method is really second order depends on how often the switches evaluate to zero, and on how the number of those points where a first order method is invoked changes as the mesh is refined. In many simulations, the number of points where the first order stencil must be chosen is relatively small, and the fraction of such points among the total often decreases as the mesh is refined. In these cases, the error is experimentally observed to be considerably reduced using this approach. The advantages of using second order operators wherever possible can be seen most clearly along diagonals, where the first order approximation to the gradient is significantly worse than the second order version. One can also use third and higher order one-sided differences; the degree to which these yield higher accuracy will again depend on how often the first order scheme is invoked. Numerical tests of the accuracy of this scheme are reported in Chapter 12. Further discussion of higher accuracy Fast Marching Methods may be found in [239].

8.9 Non-uniform orthogonal grids

The implementation of Fast Marching Methods on non-uniform orthogonal meshes is straightforward. Non-uniform Cartesian meshes are handled by the difference operators. Cylindrical and spherical coordinate meshes require the standard altered expressions for the gradient in the upwind direction. Since the use of these altered expressions does not affect flow of the Fast Marching Method, the actual algorithm remains unchanged; only the update formula changes.

8.10 General static Hamilton–Jacobi equations

Suppose we try to extend the above techniques to a more general static Hamiltonian–Jacobi equation. Thus, let $u(x)$ be a scalar function from $x = (x_1, \ldots, x_n)$: $u : R^N \to R$. The goal is to solve the static Hamilton–Jacobi equation

$$H(Du, x) = 0 \qquad (8.9)$$

where Du represents the derivatives in each of the component variables $u_{x_1}, \ldots, u_{x_N}$. As initial conditions, assume that u is known on a subset Ω of R^N (without loss of generality, assume that $u = 0$ on Ω).

Assume that H is convex. As an example, we consider a two-dimensional problem and five point stencil. (Operators with larger stencils and more space dimensions are allowed.) Suppose we approximate the Hamiltonian by a consistent difference operator of the form

$$H_{ij}(u_{ij}, u_{i-1,j}, u_{i+1,j}, u_{i,j-1}, u_{i,j+1}, x_{ij}) = 0. \qquad (8.10)$$

If this scheme allows us to write u_{ij} only in terms of smaller values of u at the neighboring points, then we can use the Fast Marching Method to systematically produce the solution. Of course, construction of such a scheme may not be straightforward, and some work may be required to devise a scheme that satisfies the above requirement (while, of course, remaining consistent and producing the viscosity solution in the sense of the previous chapter). However, if such a scheme can be generated, the Fast Marching Method can be used. One additional such scheme is given in Chapter 17.

8.11 Some clarifying comments

Both Narrow Band Level Set Methods and Fast Marching Methods require careful construction of upwind, entropy-satisfying schemes and make use of the dynamics and geometry of front propagation analyzed in [225]. However, we note that the time-dependent level set method advances the front *in unison*, while the Fast Marching Method constructs "scaffolding" to build the time solution surface T one grid point at a time. This means that the time at which the surface crosses a grid point (that is, its T value) is found before nearby values are determined. As such, there is no notion of a time step in the stationary method; one is simply constructing the stationary surface in an upwind fashion.

Finally, we revisit the issue of operation counts. Consider a computational domain in three space dimensions with N points in each grid direction. An adaptive Narrow Band Method focuses all the computational labor onto a thin band around the zero level set, thus reducing the labor to $O(N^3 k)$, where k is the width of this narrow band, providing the optimal technique for implementing level set methods. On the other hand, the Fast Marching Method is an optimal "adaptive" technique which drops the computational labor involved in solving the boundary value formulation to $O(N^3 \log N)$. At first glance, the computational efficieny of Fast Marching Methods may not be evident on the basis of these operation counts. However, two additional advantages provide the large computational savings. First, because the Narrow Band Level Set Method is solving a time-dependent problem, time step restrictions in terms of CFL conditions based on the speed F influences the number of steps required to evolve a front; in contrast, the Fast Marching Method has no such restrictions. The speed F of the front is irrelevant to the efficiency of the method. Second, the number of elements in the heap depends on the length of the front; in most cases, this length is small enough that, for all practical purposes, the sort is very fast and essentially $O(1)$.

9

Triangulated Versions of Level Set Methods

Outline: *In previous chapters, we have concentrated on methods for Cartesian orthogonal meshes. However, in some applications, triangulated unstructured mesh versions of our schemes are desired. The reasons include the versatility of meshes which are body-fit to moving interfaces, the ability to adaptively refine the mesh in regions of interest, and the ability to use irregular meshes to tessellate complex regions and manifolds. In this chapter, we develop triangulated unstructured mesh versions of Level Set Methods and general schemes for Hamilton-Jacobi equations.*[1]

While finite difference approximations on fixed, logically rectangular meshes offer high degrees of accuracy and programming ease, in some situations a triangulated unstructured mesh approximation is preferable. Reasons for desiring such a discretization include:

- **Adaptive mesh refinement:** Adaptive mesh refinement is straightforward in a triangulated setting because of the ability to subdivide elements while avoiding nonconforming approximations, i.e., "hanging nodes." Thus, in problems where one wants additional resolution, not just around the interface, but in response to other variables as well, this approach is valuable. While adaptive Cartesian level set methods have been built (see [178]), such methods have additional complications, especially in the presence of parabolic curvature terms.

- **Interface-fitted coordinates:** In some moving interface problems,

---

[1] T. Barth's contribution of unstructured mesh technologies was instrumental to the implementation of level set and Hamilton-Jacobi solvers in triangulated settings. This material is taken with little change from the paper by Barth and Sethian [24].

jump conditions across the boundary are critical for both solving partial differential equations on either side of the interface and evaluating the speed of the interface. Interpolation of these terms to neighboring grid elements can be delicate. In contrast, an interface-fitted coordinate system often offers a straightforward way to build accurate approximations to these terms. In a triangulated setting, a local interface-fitted set of nodes, similar to mesh adaptivity, can be a considerable advantage.

• **Propagating interfaces on manifolds:** In some applications, one wants to compute the motion of interfaces on non-planar manifolds such as the surface of a body. In this case, a unstructured mesh may be more readily available than an orthogonal rectilinear coordinate system.

In this chapter, we consider triangulated unstructured mesh methods for the level set equation and for the general Hamilton-Jacobi equation. We begin with the simplest possible scheme, which is a monotone update scheme. While valuable, this scheme lacks smoothness, and can be replaced by positive coefficient schemes, which are then given. This then leads to Petrov-Galerkin schemes, including those with discontinuity-capturing operators. This methodology was developed by Barth and Sethian [24]; much of this first section is taken from that reference, and we refer the interested reader to [24] (see also ([23])). This chapter assumes some familiarity with the basic ideas of finite element methods, including the ideas of basis functions and variational/energy schemes. We refer the interested reader to the excellent introduction to this field provided by C. Johnson [122].

9.1 Fundamentals and notation

We begin with the general form of the Hamilton-Jacobi equation, namely,

$$u_t + H(\nabla u) = f(x) \qquad (x,t) \in \Omega \times \mathbf{R}^+$$
$$u(x,0) = u_0(x). \tag{9.1}$$

Let $\mathcal{T}$ denote a triangulation set in $\mathbf{R}^d$, $\mathcal{T} = \{T_1, T_2, \ldots, T_{|\mathcal{T}|}\}$, composed of simplices covering Ω such that $\mathcal{T} = \cup T_j$, where the intersection of two distinct elements can only be along the common edge. We will also refer to the vertex set $V = \{v_1, v_2, \ldots, v_{|V|}\}$. Unless otherwise stated, the solution on $\mathcal{T}$ is approximated by elements in a standard piecewise

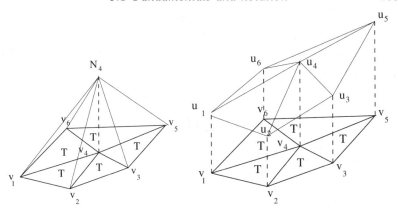

Fig. 9.1. Notation for finite elements.

linear finite element subspace denoted by $\mathcal{V}_h^1$. For $(x, t) \in \mathcal{V}_h^1 \times \mathbf{R}^+$, the numerical solution at vertex v_j at a time $n\Delta t$ is denoted by u_j^n; see Figure 9.1.

Suppose we consider a numerical approximation of Eqn. 9.1 of the form

$$u_j^{n+1} = u_j^n - \Delta t \, \mathcal{H}_j(\nabla u_1^n, \nabla u_2^n, \ldots, \nabla u_{|\mathcal{T}|}^n, x). \qquad (9.2)$$

Here, the numerical Hamiltonian is written as

$$\mathcal{H}_j(\nabla u_1, \nabla u_2, \ldots, \nabla u_{|\mathcal{T}|}, x) = \frac{\sum_{l=1}^{|\mathcal{T}|} \alpha_j^l \left(H(\nabla u) - f(x) \right)_{T_l}}{\sum_{l=1}^{|\mathcal{T}|} \alpha_j^l \, \text{meas}(T_l)}, \quad \alpha_j^l \geq 0. \qquad (9.3)$$

As written, the numerical Hamiltonian which updates the value of u at the vertex j can, in theory, use values of the gradient from all of the simplices. The contribution of each element to the update is given by the weighting coefficients $\alpha_j^l = \alpha_j(\nabla u_l)$. As will be shown, these coefficients are bounded and positive. In developing numerical discretizations, we follow Crandall and Lions [73], Osher and Sethian [187], and Abgrall [1] in proposing design criteria that reflect properties of the underlying H-J differential equation (without source term):

(i) Consistency. If $u(x, t)$ varies linearly in space and time, consistency dictates that the numerical scheme (9.2) must yield the exact evolution:

$$u(x_j, n\Delta t) = u_0(x_j) - n\Delta t \, H(\nabla u_0). \qquad (9.4)$$

In other words, a reasonable scheme must exactly approximate a linear solution. In our formulations this is always satisfied, since the assumed form of the numerical Hamiltonian (Eqn. 9.3) satisfies

$$\mathcal{H}(\nabla u, \nabla u, \dots, \nabla u) = H(\nabla u), \tag{9.5}$$

and we will produce schemes such that at least one non-zero α_j^l exists for each interior vertex v_j location.

(ii) Monotonicity. Crandall and Lions [73] have previously shown that consistent, monotone schemes for the H-J equations are stable and converge to the correct viscosity limit solution. Monotonicity, as used here, can be defined in terms of order preservation, i.e.,

$$u^n \geq v^n \qquad \text{implies} \qquad u^{n+1} \geq v^{n+1}. \tag{9.6}$$

In other words, if one set of data is larger than another set, then that ordering must be preserved for all time. One way to guarantee this monotonicity is to utilize the partial derivatives $\frac{\partial G_j}{\partial u_i}$. For self-maps of the form

$$u_j^{n+1} = u_j^n - \Delta t\, G_j(u^n), \tag{9.7}$$

it is sufficient to show that

$$\frac{\partial G_j}{\partial u_i} \leq 0 \quad \text{and} \quad 0 \leq \Delta t \leq \left(\frac{\partial G_j}{\partial u_j}\right)^{-1}, \ \forall\, i, j \in \{1, 2, \dots, |V|\}, i \neq j \tag{9.8}$$

for the scheme to be monotone.

(iii) Monotonicity will turn out to be too restrictive. Effective schemes can be obtained by satisfying a weaker *positivity* condition, namely

$$u_j^{n+1} = u_j^n - \Delta t \sum_{i=1, i \neq j}^{|V|} \beta_j^i(u^n)\,(u_j^n - u_i^n), \tag{9.9}$$

with

$$\beta_j^i \geq 0 \quad \text{and} \quad 0 \leq \Delta t \leq \left(\sum_{i=1, i \neq j}^{|V|} \beta_j^i\right)^{-1}, \quad j = 1, 2, \dots, |V|. \tag{9.10}$$

This is a statement about the scheme itself (that is, the effect of β_j^i), rather than a statement about the partials of G. Both the

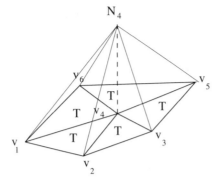

Fig. 9.2. Basis element.

monotone and positive schemes satisfy a global discrete maximum principle, namely

$$\min_{i=1,2,\dots,|V|} u_i^n \le u_j^{n+1} \le \max_{i=1,2,\dots,|V|} u_i^n, \quad \forall\, j = 1, 2, \dots, |V|,$$
(9.11)

as well as the local maximum principle

$$\min_{i\in\text{supp}_j} u_i^n \le u_j^{n+1} \le \max_{i\in\text{supp}_j} u_i^n, \quad \forall\, j = 1, 2, \dots, |V|, \qquad (9.12)$$

where supp_j denotes the index support set for the discretized scheme at vertex v_j.

Our goal is to produce procedures for calculating the α_j^l coefficients so that monotonicity and/or positivity properties can be obtained for specific Hamiltonians.

9.2 A monotone scheme for $H(\nabla u)$

We begin with a very straightforward scheme. Consider a d-dimensional simplex T with linearly varying $u(x)$ uniquely defined by vertex values u_i and linear basis functions $N_i(x)$ defined by $N_i(x_j) = \delta_{ij}$, where δ_{ij} is zero except when $i = j$ (see Figure 9.2).

$$u(x) = \sum_{i=1}^{d+1} N_i(x)\, u_i. \qquad (9.13)$$

The gradient ∇u in simplex T is readily computed in terms of gradients of the basis functions

$$\nabla u = \sum_{i=1}^{d+1} \nabla N_i \, u_i. \tag{9.14}$$

To gain a better geometric intuition, the gradient formula is rewritten in terms of inward pointing normals $\vec{n}_i$ scaled by the measure of the $(d-1)$-facet opposite vertex v_i in the simplex, i.e.,

$$\nabla u = \frac{1}{d \, \mathrm{meas}(T)} \sum_{i=1}^{d+1} \vec{n}_i \, u_i. \tag{9.15}$$

Note that

$$\sum_{i=1}^{d+1} \vec{n}_i = 0, \tag{9.16}$$

because of the scaling of normals and the fact that the surface of a simplex is closed. We can now define the gradient of the Hamiltonian with respect to the gradient arguments

$$\nabla H \equiv \begin{pmatrix} \frac{\partial H}{\partial u_x} \\ \frac{\partial H}{\partial u_y} \\ \vdots \end{pmatrix}. \tag{9.17}$$

Next, form $\left(H(\nabla u) \right)_T = \int_T H(\nabla u) \, d\Omega$, which is the Hamiltonian integrated in a simplex, and differentiate with respect to the simplex unknowns, thus

$$\frac{\partial \left(H(\nabla u) \right)_T}{\partial u_i} = \frac{1}{d} (\nabla H \cdot \vec{n}_i), \quad i = 1, 2, \ldots, d+1. \tag{9.18}$$

Because of the scaling of normals and the constancy of $H(\nabla u)$ within a linear simplex, we must have that

$$\sum_{i=1}^{d+1} \frac{\partial \left(H(\nabla u) \right)_T}{\partial u_i} = 0. \tag{9.19}$$

In order to enforce the monotonicity conditions (Eqn. 9.8), we need to consider the orientations of ∇H relative to the geometry of simplex T. Figure 9.3 shows the correct way to maintain monotonicity for a single isolated two-dimensional simplex. The various cases produce the following:

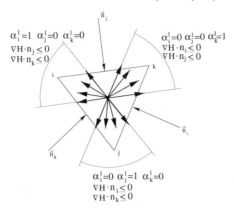

Fig. 9.3. Typical 2-D simplex T_l showing inward pointing normals, index convention, and sector demarcation.

$\nabla H \cdot n_j \leq 0, \nabla H \cdot n_k \leq 0$:

$$u_i^{n+1} = u_i^n - \Delta t \, \frac{\left(H(\nabla u^n)\right)_T}{\text{meas}(T)}, \qquad (9.20)$$

$\nabla H \cdot n_i \leq 0, \nabla H \cdot n_k \leq 0$:

$$u_j^{n+1} = u_j^n - \Delta t \, \frac{\left(H(\nabla u^n)\right)_T}{\text{meas}(T)}, \qquad (9.21)$$

$\nabla H \cdot n_i \leq 0, \nabla H \cdot n_j \leq 0$:

$$u_k^{n+1} = u_k^n - \Delta t \, \frac{\left(H(\nabla u^n)\right)_T}{\text{meas}(T)}. \qquad (9.22)$$

This construction is equivalent to the idea of upwinding and accounts for the correct direction of propagation of information. In the more general case in which several simplices contribute to the update at a single vertex, the generalized numerical Hamiltonian formula (Eqn. 9.3) can be used, namely

$$u_j^{n+1} = u_j^n - \Delta t \, \frac{\sum_{l=1}^{|\mathcal{T}|} \alpha_j^l \left(H(\nabla u^n)\right)_{T_l}}{\sum_{l=1}^{|\mathcal{T}|} \alpha_j^l \, \text{meas}(T_l)}, \qquad (9.23)$$

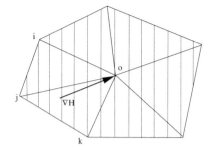

Fig. 9.4. Triangles surrounding v_0 with linearly varying $u(x)$. Iso-levels of constant $u(x)$ are shown as straight parallel lines and the arrow denotes the constant gradient vector ∇H at vertex v_0.

where

$$\alpha_j^l = \begin{cases} 1 & \nabla H \cdot n_k \leq 0, \ k = 1, 2, \ldots, d+1, j \neq k \\ 0 & \text{otherwise} \end{cases}, \qquad x \in T_l.$$

$$(9.24)$$

The scheme is consistent if, given the form of the numerical Hamiltonian, it can be shown that there exists at least one simplex producing a nonzero α_j^l coefficient when $u(x)$ varies linearly over the entire support of the scheme. From Figure 9.4, it is clear that this must always happen. Consider a constant gradient vector ∇H located at a vertex v_0 to be updated. By simply tracing backward along the gradient vector into a simplex T_m surrounding v_0, this must lead to a simplex which contributes a nonzero α_0^m.

As straightforward as this scheme is, it suffers from a flaw. The numerical Hamiltonian fails to be a Lipschitz continuous function and varies discontinuously with certain small perturbations in its arguments. This can greatly reduce the accuracy of solutions. In [24], a virtual edge flipping strategy was developed which uses adjacent information to re-orient the diagonal and hence ensures that information comes without this discontinuous jump. We note two factors about this edge-flipping strategy:

- It increases the support of the update of a point and hence requires more computation per update.
- The edge flipping requires that adjacent triangle pairs form convex quadrilaterals, which means that the underlying triangulation must be acute.

Nonetheless, this monotone scheme is a straightforward approach to solving the Hamilton-Jacobi equation.

9.3 A positive scheme for homogeneous $H(\nabla u)$

To gain Lipschitz continuity of the numerical Hamiltonian without resorting to virtual edge flipping and hence preserving the locality of an update, the condition of monotonicity is relaxed in favor of a positivity condition. In this new framework, we want to show that the $d + 1$ coefficients α_j^l for the simplex T_l are continuous functions of the gradient vector ∇u. For accuracy reasons α_j^l should also be bounded and positive. Note that since monotone schemes depend fundamentally on the quantity

$$\frac{\partial \left(H(\nabla u) \right)_T}{\partial u_i} = \frac{1}{d} \nabla H \cdot \vec{n}_i,$$

any new positive scheme should, in principle, also depend on this quantity. This would ensure that monotone updates would be recovered whenever possible.

Recall Euler's theorem for homogeneous functions: A function $F(u, v) :$ $\mathbf{R} \times \mathbf{R} \to \mathbf{R}$ is homogeneous of degree p if $F(tu, tv) = t^p F(u, v)$, and hence we must have that

$$pF(u, v) = \frac{\partial F}{\partial u} u + \frac{\partial F}{\partial v} v. \tag{9.25}$$

We now focus attention on Hamiltonians $H(\nabla u)$ which are homogeneous functions of degree p in ∇u. In this case, Euler's theorem yields the following relation for a simplex T in $\mathbf{R}^d$:

$$\left(H(\nabla u) \right)_T = \frac{1}{p} \nabla H \cdot \nabla u = \frac{1}{d\,p} \sum_{i=1}^{d+1} (\nabla H \cdot \vec{n}_i) u_i. \tag{9.26}$$

Next, define

$$K_i = \frac{\nabla H \cdot \vec{n}_i}{d\,p}, \tag{9.27}$$

so that Eqn. 9.26 can be expressed in the following canonical form:

$$\left(H(\nabla u) \right)_T = \sum_{i=1}^{d+1} K_i u_i, \quad \text{with} \quad \sum_{i=1}^{d+1} K_i = 0. \tag{9.28}$$

We can now draw upon a number of well-known techniques for constructing positive coefficient schemes for conservation laws due to Roe

[204] [205] and Deconinck et al. [77]. The basis for these schemes is the following straightforward manipulation of the canonical form:

$$
\begin{aligned}
\sum_{i=1}^{d+1} K_i u_i &= \sum_{j=1}^{d+1} K_j^- u_j + \sum_{i=1}^{d+1} K_i^+ u_i \\
&= \left(\sum_{l=1}^{d+1} K_l^+ \right)^{-1} \left(\sum_{i=1}^{d+1} K_i^+ \right) \sum_{j=1}^{d+1} K_j^- u_j \\
&\quad - \left(\sum_{l=1}^{d+1} K_l^+ \right)^{-1} \left(\sum_{j=1}^{d+1} K_j^- \right) \sum_{i=1}^{d+1} K_i^+ u_i \\
&= \sum_{i=1}^{d+1} K_i^+ \left(\sum_{l=1}^{d+1} K_l^+ \right)^{-1} \sum_{j=1}^{d+1} K_j^- (u_j - u_i). \quad (9.29)
\end{aligned}
$$

In deriving this formula, the identity $\sum_{j=1}^{d+1} K_j = \sum_{j=1}^{d+1} K_j^+ + \sum_{j=1}^{d+1} K_j^- = 0$ has been used. Eqn. 9.29 motivates a decomposition of the Hamiltonian in a simplex T of the form

$$
\left(H(\nabla u) \right)_T = \sum_{i=1}^{d+1} \delta_i, \quad (9.30)
$$

with

$$
\delta_i = K_i^+ \left(\sum_{l=1}^{d+1} K_l^+ \right)^{-1} \sum_{j=1}^{d+1} K_j^- (u_j - u_i). \quad (9.31)
$$

When written in matrix form,

$$
\begin{pmatrix} \delta_1 \\ \delta_2 \\ \vdots \\ \delta_{d+1} \end{pmatrix} = \begin{bmatrix} \ddots & & - \\ & + & \\ - & & \ddots \end{bmatrix} \begin{pmatrix} u_1 \\ u_2 \\ \vdots \\ u_{d+1} \end{pmatrix}, \quad (9.32)
$$

the decomposition reveals a sign pattern useful in constructing a positive coefficient schemes satisfying Eqn. (9.9). To do so, simply let

$$
\alpha_l^i = \frac{\delta_i}{\left(H(\nabla u) \right)_{T_l}}, \quad (9.33)
$$

and insert into the prototype numerical Hamiltonian

$$\mathcal{H}_j(\nabla u_1, \nabla u_2, \ldots, \nabla u_{|\mathcal{T}|}) = \frac{\sum_{l=1}^{|\mathcal{T}|} \alpha_j^l \left(H(\nabla u)\right)_{T_l}}{\sum_{l=1}^{|\mathcal{T}|} \alpha_j^l \, \text{meas}(T_l)} = \frac{\sum_{l=1}^{|\mathcal{T}|} \delta_j^l}{\sum_{l=1}^{|\mathcal{T}|} \alpha_j^l \, \text{meas}(T_l)}. \tag{9.34}$$

Since the coefficient terms depend continuously on ∇H and the simplex geometry, Lipschitz continuity of the numerical Hamiltonian is obtained. Note that, by construction, $\sum_{j=1}^{d+1} \alpha_j^l = 1$. It is not difficult to show that for certain Hamiltonians (e.g. $H(\nabla u) = |\nabla u|^p$) and *non-obtuse* triangulations, α_j^l are nonnegative and hence bounded. In this case the conditions for a positive coefficient scheme are obtained:

$$u_j^{n+1} = u_j^n - \Delta t \sum_{i=1, i\neq j}^{|V|} \beta_j^i(u^n) \, (u_j^n - u_i^n), \tag{9.35}$$

with

$$\beta_j^i \geq 0 \quad \text{and} \quad 0 \leq \Delta t \leq \left(\sum_{i=1, i\neq j}^{|V|} \beta_j^i\right)^{-1}, \quad j = 1, 2, \ldots, |V|. \tag{9.36}$$

For general (obtuse) triangulations and homogeneous Hamiltonians, additional nonlinearity must be introduced into the formulation so that a positive coefficient scheme is obtained. Deconinck et. al. [77], in the context of conservation law equation discretizations, suggest a simple nonlinear modification of α_j^l to obtain a positive coefficient scheme for general triangulations, namely

$$\tilde{\alpha}_j^l = \frac{\max(0, \alpha_j^l)}{\sum_{k=1}^{d+1} \max(0, \alpha_k^l)}. \tag{9.37}$$

The inclusion of this modification produces a robust scheme for the H-J equations (with source term), i.e.,

$$\mathcal{H}_j(\nabla u_1, \nabla u_2, \ldots, \nabla u_{|\mathcal{T}|}, x) = \frac{\sum_{l=1}^{|\mathcal{T}|} \tilde{\alpha}_j^l \left(H(\nabla u) - f(x)\right)_{T_l}}{\sum_{l=1}^{|\mathcal{T}|} \tilde{\alpha}_j^l \, \text{meas}(T_l)}. \tag{9.38}$$

This nonlinear modification still allows the scheme to be written in the positive coefficient form (9.35).

9.4 A Petrov-Galerkin formulation

Finally, we consider one more scheme, which is a stabilized finite element approximation for the Hamilton-Jacobi equation

$$u_t + H(\nabla u) = 0. \tag{9.39}$$

Again, the Hamiltonian $H(\nabla u)$ is assumed to be homogeneous of degree p, that is,

$$H(\nabla u) = p^{-1}\, \nabla H \cdot \nabla u \tag{9.40}$$

for positive p. The formulation considered here is inspired by the stabilized Petrov-Galerkin methods developed for conservation law equations (see Johnson [122] for a detailed discussion). Let $\mathcal{P}_k$ denote k-th order interpolation polynomials in each simplex T and $\mathcal{D}^n = \Omega \times I^n$ the space-time slab with $I^n = [t^n_+, t^{n+1}_-]$. For a given positive integer k define the usual trial space

$$\mathcal{V}^h_n = \{v^h | v^h \in C^0(\mathcal{D}^n), v^h|_{T \times I^n} \in \mathcal{P}_k(T \times I^n)\ \forall T \in \mathcal{T}\}, \tag{9.41}$$

i.e. $\mathcal{V}^h$ is the space of continuous piecewise polynomial functions of degree k for the space-time slab $\mathcal{D}^n$. Note that between space-time slabs, $\mathcal{V}^n$ is allowed to jump discontinuously, i.e. $u^h(x, t^n_-) \neq u^h(x, t^n_+)$. Next define the inflow/outflow portions of the boundary Γ:

$$\Gamma_+ = \Gamma \setminus \Gamma_- = \{x \in \Gamma : n(x) \cdot \nabla H \geq 0\}. \tag{9.42}$$

Now, the following stabilized finite element approximation with weakly imposed boundary conditions can be stated:

Finite Element Formulation:

Find $u^h \in \mathcal{V}^h_n$ such that for all $w^h \in \mathcal{V}^h_n$

$$B(u^h, w^h)_{gal} + B(u^h, w^h)_{ls} + B(u^h, w^h)_{bc} = 0, \tag{9.43}$$

with

$$B(u, w)_{gal} = \int_{I^n} \int_\Omega w\, (u_t + H(\nabla u))\ d\Omega\, dt,$$

$$B(u, w)_{ls} = \int_{I^n} \int_\Omega (w_t + p^{-1}(\nabla H \cdot \nabla w))\, \tau\, (u_t + p^{-1}(\nabla H \cdot \nabla u))\ d\Omega\, dt,$$

$$B(u, w)_{bc} = \int_{I^n} \int_{\Gamma_-} w\, (g - u)\, p^{-1}\, (\nabla H \cdot n)\ d\Gamma\, dt$$

$$+ \int_\Omega w(t^n_+)(u(t^n_+) - u(t^n_-))\ d\Omega, \tag{9.44}$$

for $\tau > 0$ and $\nabla H = \nabla H(\nabla u)$ everywhere in (9.44).

The motivation for the least-squares stabilization $B(u,w)_{ls}$ comes from looking at the variation of the quadratic potential

$$\mathcal{P}(u) = \int_{\mathcal{D}^n} \frac{1}{2}\tau \left(u_t + p^{-1}(\nabla H \cdot \nabla u)\right)^2 d\mathcal{D} \qquad (9.45)$$

for constant ∇H where

$$\delta \mathcal{P} = \lim_{\sigma \to 0} \frac{d}{d\sigma}\mathcal{P}(u + \sigma w) = B(u,w)_{ls}, \quad \forall w \in \mathcal{V}. \qquad (9.46)$$

It can be shown that the least squares term removes energy from the system. However, unlike monotone and positive coefficient discretizations, this is not enough to control oscillations in the solution. Consequently, small solution oscillations are sometimes present near slope discontinuities. To remove these oscillations, a discontinuity-capturing operator is employed that is similar to the one commonly used for conservation law discretizations:

$$B(u,w)_{dc} = \int_T \nu(u)(\nabla w \cdot \nabla u)\, dx, \quad \nu = \frac{\sigma_{dc}\, h\, |u_t + H(\nabla u)|}{(|u_t|^q + |H(\nabla u)|^q + \epsilon^q)^{1/q}}. \qquad (9.47)$$

In this formula, ϵ is a small computer dependent parameter, and q is a parameter with typical values $q \in \{1, 2, \infty\}$. This form of the discontinuity capturing can be viewed as a form of non-linear artificial viscosity. Finally, a reasonable form for the parameter τ is

$$\tau = \left(\left(\frac{2}{\Delta t}\right)^2 + \left(\frac{2|\nabla H|}{ph}\right)^2\right)^{-1/2}. \qquad (9.48)$$

For details, see [24].

9.5 Time integration schemes

The Hamilton-Jacobi and level set equations both require accurate time integration schemes. In a relatively straightforward manner, single- and two-stage Runge-Kutta time advancement schemes can be built for the monotone and positive schemes (see below). For the Petrov-Galerkin schemes, the formulation permits piecewise constant in time elements. Higher order accuracy in time is naturally achieved using piecewise discontinuous polynomial approximations in time.

9.6 Algorithms

For the sake of completeness, we include below the implementation of each scheme for the level set equation. Again, we refer the interested reader to [24], where further schemes are laid out. These include schemes which focus on a space-time formulation, view time as an additional space-like variable, and converge to a solution of the space-time formulation.

9.6.1 *The explicit positive coefficient scheme*

Consider the level set equation

$$\phi_t + F(\nabla\phi, x)|\nabla\phi| = f(x), \quad x \in \mathbf{R}^d \tag{9.49}$$

discretized using the positive coefficient scheme described in Sec. 9.3. This algorithm can be implemented using the following algorithmic steps:

(1) Initialize $\phi_i^* = w_i = 0$, $i = 1, 2, \ldots, |V|$.

(2) For each $T \in \mathcal{T}$, $i = 1, 2, \ldots, d+1$

$$T \overset{local}{\longrightarrow} \text{simplex}(x_1, x_2, \ldots, x_{d+1})$$
$$N_i(x) = \{N_i(x) \in \mathcal{P}_1 \mid N_i(x_j) = \delta_{ij}, \ j = 1, 2, \ldots, d+1, \ x \in T\}$$
$$\overline{F} = \frac{1}{\text{meas}(T)} \int_T F(\nabla\phi) \, dx$$
$$\overline{f} = \frac{1}{\text{meas}(T)} \int_T f(x) \, dx$$
$$\vec{n}_i = d \, \text{meas}(T) \, \nabla N_i$$
$$\nabla\phi = \sum_{j=1}^{d+1} \nabla N_j \, \phi_j$$
$$K_i = \frac{\overline{F} \, \nabla\phi \cdot \vec{n}_i}{d \, |\nabla\phi|}$$
$$\delta\phi = \sum_{l=1}^{d+1} K_l \, \phi_l$$
$$\delta\phi_i = K_i^+ \left(\sum_{l=1}^{d+1} K_l^-\right)^{-1} \sum_{l=1}^{d+1} K_l^- (\phi_i - \phi_l)$$
$$\tilde{\alpha}_i = \frac{\max(0, \delta\phi_i/\delta\phi)}{\sum_{l=1}^{d+1} \max(0, \delta\phi_l/\delta\phi)}$$
$$\phi_i^* = \phi^* + \tilde{\alpha}_i \, (\delta\phi - \overline{f} \, \text{meas}(T))$$
$$w_i = w_i + \tilde{\alpha}_i \, \text{meas}(T)$$

End for

Single-stage integration

(3) For each $v_i \in V$

$$\phi_i^{n+1} = \phi_i^n - \Delta t \, \frac{(\phi_i^*)^n}{w_i^n}$$

EndFor

Two-stage integration

(3) For each $v_i \in V$

$$\phi_i^{n+1/2} = \phi_i^n - \Delta t \, \frac{(\phi_i^*)^n}{w_i^n}$$

$$\phi_i^{n+1} = \frac{1}{2}(\phi_i^n + \phi_i^{n+1/2}) - \frac{\Delta t}{2} \, \frac{(\phi_i^*)^{n+1/2}}{w_i^{n+1/2}}$$

EndFor

9.6.2 The explicit Petrov-Galerkin scheme

Once again consider the level set equation

$$\phi_t + F(\nabla\phi, x)|\nabla\phi| = f(x), \quad x \in \mathbf{R}^d. \tag{9.50}$$

The Petrov-Galerkin implementation follows closely the implementation for the positive coefficient scheme:

(1) Initialize $\phi_i^* = w_i = 0$, $i = 1, 2, \ldots, |V|$.

(2) For each $T \in \mathcal{T}$, $i = 1, 2, \ldots, d+1$

$$T \xrightarrow{local} \text{simplex}(x_1, x_2, \ldots, x_{d+1})$$
$$N_i(x) = \{N_i(x) \in \mathcal{P}_1 \mid N_i(x_j) = \delta_{ij}, \; j = 1, 2, \ldots, d+1, \; x \in T\}$$
$$\overline{F} = \frac{1}{\text{meas}(T)} \int_T F(\nabla\phi) \, dx$$
$$\overline{f} = \frac{1}{\text{meas}(T)} \int_T f(x) \, dx$$
$$\vec{n}_i = d \, \text{meas}(T) \, \nabla N_i$$
$$\nabla\phi = \sum_{j=1}^{d+1} \nabla N_j \, \phi_j$$
$$K_i = \frac{\overline{F} \, \nabla\phi \cdot \vec{n}_i}{d \, |\nabla\phi|}$$
$$\tau = \tau(\phi)$$
$$\delta\phi = \sum_{l=1}^{d+1} K_l \, \phi_l$$
$$\alpha_i = \frac{1}{d} + \tau \, K_i$$
$$\phi_i^* = \phi^* + \alpha_i \, (\delta\phi - \overline{f} \, \text{meas}(T))$$
$$w_i = w_i + \alpha_i \, \text{meas}(T)$$

End for

Single-stage integration

(3) For each $v_i \in V$

$$\phi_i^{n+1} = \phi_i^n - \Delta t \, \frac{(\phi_i^*)^n}{w_i^n}$$

EndFor
Two-stage integration

(3) For each $v_i \in V$

$$\phi_i^{n+1/2} = \phi_i^n - \Delta t \, \frac{(\phi_i^*)^n}{w_i^n}$$

$$\phi_i^{n+1} = \tfrac{1}{2}(\phi_i^n + \phi_i^{n+1/2}) - \frac{\Delta t}{2} \, \frac{(\phi_i^*)^{n+1/2}}{w_i^{n+1/2}}$$

EndFor

9.7 Schemes for curvature flow

As discussed earlier, many physically relevant problems require a second derivative curvature flow term $\kappa \, |\nabla \phi|$ for the level set equation

$$\phi_t + (F - \epsilon \kappa) \, |\nabla \phi| = 0, \tag{9.51}$$

for locally constant ϵ. Recall that κ, the mean curvature of the level set function ϕ, is given by

$$\kappa = \nabla \cdot \frac{\nabla \phi}{|\nabla \phi|}. \tag{9.52}$$

In discretizing the curvature term, we follow our previous reasoning that the numerical stability will be dictated by the highest order differential term. This motivates a study of the following model equation on $\Omega \subset \mathbf{R}^2$ with boundary Γ:

$$
\begin{aligned}
\phi_t - \epsilon \nabla \cdot \frac{\nabla \phi}{|\nabla \phi|} &= 0, & x \in \Omega \\
\phi(x,0) &= \phi_0(x), & x \in \Omega \\
\phi(x,t) &= g(x), & x \in \Gamma.
\end{aligned}
\tag{9.53}
$$

This corresponds to flow under curvature, which can now be restated in variational form. Let $\mathcal{S}^h$ denote the space of finite-dimensional functions on the grid with bounded energy, $\int_\Omega |\nabla u| \, dx$, satisfying the Dirichlet boundary condition and $\mathcal{V}^h$ the same space of finite-dimensional functions with bounded energy which satisfy homogeneous boundary conditions. Given these spaces, the variational goal is to find $\phi \in \mathcal{S}^h$, such that for all $w \in \mathcal{V}^h$,

$$
\begin{aligned}
\int_\Omega w u_t \, dx + \epsilon \int_\Omega \frac{\nabla w \cdot \nabla \phi}{|\nabla \phi|} \, dx &= 0 \\
\phi(x,0) &= \phi_0(x), & x \in \Omega \\
\phi(x,t) &= g(x), & x \in \Gamma.
\end{aligned}
\tag{9.54}
$$

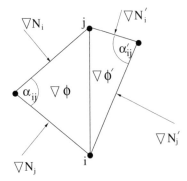

Fig. 9.5. Discretization weight geometry for the edge $e(v_i, v_j)$.

Examining Eqn. 9.54 with linear elements, and with some manipulation, one can produce weights of the form (see Figure 9.5)

$$
\begin{aligned}
W_j^i &= \left[\frac{\nabla N_i \cdot \nabla N_j}{|\nabla \phi|} + \frac{\nabla N_i' \cdot \nabla N_j'}{|\nabla \phi|'} \right] \\
&= \frac{1}{2} \left[\frac{\cotan(\alpha_{ij})}{|\nabla \phi|} + \frac{\cotan(\alpha_{ij}')}{|\nabla \phi|'} \right],
\end{aligned}
\tag{9.55}
$$

where N_i denotes the set of vertices adjacent to vertex v_i.

Thus, the strategy for the general level set curvature flow term

$$
|\nabla \phi| \, \nabla \cdot \frac{\nabla \phi}{|\nabla \phi|}
\tag{9.56}
$$

is to obtain a pointwise estimate at each vertex v_i of the form

$$
|\nabla \phi| \, \nabla \cdot \frac{\nabla \phi}{|\nabla \phi|} \bigg|_i \approx |\nabla \phi|_i \, \frac{\sum_{j \in \mathcal{N}_i} W_j^i \, (u_j - u_i)}{\sum_{T \in N_i} \mathrm{meas}(T)},
\tag{9.57}
$$

where W_j^i are the weights described above, N_i denotes the triangle neighbor set incident to v_i, and $(\nabla \phi)_i$ is the pointwise lumped-Galerkin approximation, i.e.

$$
(\nabla \phi)_i = \frac{\sum_{T \in N_i} \int_T \nabla u \, dx}{\sum_{T \in N_i} \mathrm{meas}(T)}.
\tag{9.58}
$$

This term is easily incorporated into the previously discussed explicit schemes. For details, see [24].

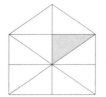 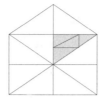

(a) Candidate triangle   (b) Conformal subdivision   (c) Subdivision resolution

Fig. 9.6. Conformal refinement steps for a 2D triangle.

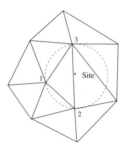 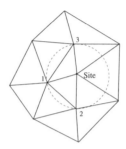

Fig. 9.7. Steiner refinement: site insertion at circumcenter location (left) and edge reconfiguration (right).

9.8 Mesh adaptivity

One of the central virtues of a triangulated domain formulation, as opposed to a rectilinear framework, is the ease in implementing mesh adaptivity. Mesh adaptation either can be part of the initial mesh construction, or can occur dynamically as the solution evolves. Two basic adaptation strategies are:

- Conformal adaptation: The term "conformal adaptation" comes from the fact that a two-dimensional triangle can be divided into four subtriangles, each having the same angles as the original triangle. Using this idea, a natural two-step algorithm exists to sub-divide any triangle; first, the selected triangle is conformally sub-divided by connecting the midpoints of each side, and then these side "hanging nodes" are then resolved by connecting them to the far vertex. The result is again a triangulated grid with no hanging nodes. The technique extends to three dimensions (see for example [154]), although the method ceases to preserve angles.

- Steiner/Delaunay adaptation: The Steiner/Delaunay adaptation technique exploits the well-known circumcircle/circumsphere characterization of unconstrained Delaunay triangulations, i.e., that the circum-circle/circumsphere of any triangle/tetrahedron does not contain any other vertex in the triangulation. For example in Figure 9.7, the triangle $T(v_1, v_2, v_3)$ is refined by adding a site at the circumcenter location and reconfiguring edges using an edge flipping procedure to restore the Delaunay characterization; see [58], [209].

In Chapter 12, we analyze the performance of these triangulated schemes on some test problems.

10

Triangulated Fast Marching Methods

Outline: *In this chapter, we extend our algorithms for unstructured meshes and develop Fast Marching Methods for triangulated domains. This is done in three stages. First, we reinterpret the Cartesian Fast Marching Method presented earlier as an algorithm on a regular triangular grid formed by cutting each square into triangles. Using this as motivation, we then extend this scheme in a natural way to provide an upwind causality scheme for acute triangulations. We then end by showing one way to use this methodology on non-acute meshes.*

Now that we have built versions of our basic schemes for solving Hamilton-Jacobi equations and Level Set Methods on triangulated meshes, the natural next step is to extend these ideas to Fast Marching Methods. A triangulated Fast Marching Method for solving the Eikonal and related equations is particularly useful for computing geodesic shortest paths on manifolds and predicting photolithography development in non-planar regions. The triangulated version of Fast Marching Methods was first developed by Kimmel and Sethian [137], using the unstructured mesh methodology of Barth and Sethian [24], and applied to the problem of constructing geodesic shortest paths on manifolds. Various examples were given there, and we follow that work closely in this presentation. For further details, see [137].

10.1 The update procedure

As an introduction, recall the update procedure in the Fast Marching Method for readjusting the values of neighbors which are downwind to known points. This is the procedure by which new trial values are created for T in the heap of nearby points. Imagine a uniform square

120

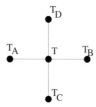

Fig. 10.1. Construction of upwind solution on orthogonal mesh.

grid and suppose that the goal is to update the value of T at the center point (i, j). We label the values of T at the surrounding grid points $T_A = T_{i-1,j}$, $T_B = T_{i+1,j}$, $T_C = T_{i,j-1}$, and $T_D = T_{i,j+1}$ (see Figure 10.1). Some of the values may be infinite, corresponding to *Far* values.

Standing at the center point, the Fast Marching Method attempts to solve the quadratic equation given by each quadrant. For example, we refer to possible contributors A and C. Without loss of generality, there are two cases:

(i) If only T_A is known, then we find the solution $T < T_A$ to the "quadratic" equation

$$(T - T_A)^2 = h^2 \frac{1}{F_{ij}^2},$$

where h is the uniform grid spacing.

(ii) If T_A and T_C are known, then we take the real solution $T \geq T_A, T_C$ to the quadratic equation

$$(T - T_A)^2 + (T - T_C)^2 = h^2 \frac{1}{F_{ij}^2},$$

For each possible up-down/left-right pair, we construct all possible real solutions; we then accept as the updated point the one that produces the smallest value of T. This is the Fast Marching Method described in [233, 235].

10.2 A scheme for a particular triangulated domain

We now to extend this method to triangular domains. In order to do so, we shall build a monotone update procedure on the triangulated mesh. As motivation, we consider the obvious triangulation of a square grid in the plane, and imagine the triangulation given in Figure 10.2.

If we consider the values T_A and T_C as *Known*, and look at the

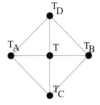

Fig. 10.2. Simple triangulation for building a monotone update operator.

triangle formed by the points which hold T_A, T_C and T, we can easily write down the equation of the plane determined by the known values T_A and T_C, as well as the unknown value T, namely (with h as the side length)

$$\left[\frac{T - T_A}{h}\right] x + \left[\frac{T - T_C}{h}\right] y + T = z.$$

Computing the gradient, we then want to select a value of T such that

$$\left[\frac{T - T_A}{h}\right]^2 + \left[\frac{T - T_C}{h}\right]^2 = \frac{1}{F_{ij}^2}.$$

In other words, we are lifting the plane by a value $T \geq T_A, T_C$ at the center point i, j in order to have a gradient magnitude equal to $1/F$. We note that this is the exact same construction as the one produced by the "orthogonal" construction. Note also that in this case the gradient vector, with origin at the center point, always points into the triangle from which it is updated. This is not necessarily the case for an arbitrary acute triangle. In order to establish monotonicity we will need to verify this condition.

The inability to solve this quadratic corresponds to an inability to tilt at an appropriate angle, and the requirement that the solution T be greater than the contributors means that the solution is always constructed in an upwind manner. Using the same update rules as before and the heap structure to maintain a list of *Trial* points, this provides a method for executing the Fast Marching Method on this simple triangulation.

Fig. 10.3. Acute triangulation around center grid point.

10.3 Fast Marching Methods on triangulated domains

Following the construction unstructured mesh upwind approximations to the gradient ([24]), we now to extend this idea to an arbitrary triangulation.

10.3.1 A construction for acute triangulations

We start with an acute triangulation and consider the triangulation around the grid point given in Figure 10.3.

A large number of triangles may share the center vertex. Our procedure, motivated by the simple triangulation in the previous section, is to compute a possible value for T from each triangle that includes the center point as a vertex. Since several triangles can produce admissible values for T, we must select an appropriate value. There are several possibilities. We chose the one that produces the smallest new value for T; this will correspond to an algorithm similar to the one used on the triangular mesh. More elaborate upwind constructions on triangulated meshes are given in [24].

Consider the non-obtuse triangle ABC in which the point to update is C. Assume that $T(B) > T(A)$. We first verify that the update is from within the triangle, i.e., the altitude h should be inside the non-obtuse triangle CBD (see Figure 10.4). This means that we search for $t = EC$ such that

$$\frac{t - u}{h} = F.$$

Denote $a = BC$ and $b = AC$; we have, by similarity, that $t/b = DF/AD = u/AD$, thus $CD = b - AD = b - bu/t = b(t-u)/u$. Next, by the Law of Cosines: $BD^2 = a^2 + CD^2 - 2a\,CD\cos\theta$, and by the Law of Sines: $\sin\phi = \frac{CD}{BD}\sin\theta$. Now, using the right angle triangle CBG we

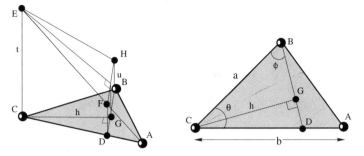

Fig. 10.4. Given the triangle ABC, such that $u = T(B) - T(A)$, find $T(C) = T(A) + t$ such that $(t - u)/h = F$. Left: a perspective view of the triangle stencil supporting the $T()$ values that form a lifted plane with a gradient magnitude equal to F. Right: the trigonometry on the plane defined by the triangle stencil.

have

$$h = a \sin \phi = a \frac{CD}{BD} \sin \theta = \frac{a\,CD\,\sin \theta}{\sqrt{a^2 + CD^2 - 2a\,CD \cos \theta}}.$$

This yields a quadratic equation for t:

$$(a^2 + b^2 - 2ab \cos \theta)t^2 + 2bu(a \cos \theta - b)t + b^2(u^2 - F^2 a^2 \sin^2 \theta) = 0.$$

The solution t must satisfy $u < t$, and should be updated from within the triangle, namely:

$$a \cos \theta < \frac{b(t - u)}{t} < \frac{a}{\cos \theta}. \qquad (10.1)$$

Thus, the update procedure is given as follows:

If $u < t$ **and** $a \cos \theta < \frac{b(t-u)}{t} < \frac{a}{\cos \theta}$,

then $T(C) = \min\{T(C), t + T(A)\}$;

else $T(C) = \min\{T(C), bF + T(A), cF + T(B)\}$.

This is our scheme to extend the Fast Marching Method to acute triangulated domains.

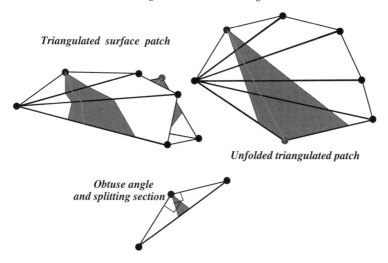

Fig. 10.5. Top: a triangulated surface patch. Bottom: the initialization of the construction for the splitting section. Right: the unfolded patch and the splitting section expansion up to the first vertex B, and the virtual edge connecting the two vertices AB.

10.3.2 Extension to general triangulations

Finally, we note that we have required an acute triangulation. This is so that any front entering the side of a triangle will have two points to provide values before the third is computed. In other words, for monotonicity, we restrict the update to come from within the triangle, i.e., the gradient of the solution at a grid point should point into the triangle from which it is updated. The most straightforward way to enforce this requirement is to build a non-obtuse triangulation, thus making sure that the grid captures all incoming fronts.

One approach to handle non-acute triangulations, which we now describe, is to build numerical support locally at obtuse angles by splitting these angles in a special way. An obtuse angle at vertex A can be updated by its neighboring points in a consistent way only at a limited section of upcoming fronts. Connecting the vertex to any point in this section splits the obtuse angle into two acute ones.

The difficulty with this approach is that we need to reach back and use more triangles than simply the one containing the point to be updated. These other triangles might not be coplanar, since the unstructured mesh may lie on a convoluted surface. The idea is to extend this section by recursively unfolding the adjacent triangle(s), until a new vertex B

is included in the extended section. Then the vertices are connected by a virtual directional edge from B to A (i.e., A may be updated by B). The length of the edge AB is equal to the distance between A and B on the unfolded triangles plane.

Finally, we note that a complexity analysis for this unfolding procedure was given in [137], showing that the $O(N \log N)$ operation count is maintained. We refer the interested reader there for further details. Further discussion of triangulated Fast Marching Methods, higher order versions, and extensions to the basic techniques may be found in [244].

11

Coupling to Applications: The Fast Construction of Extension Velocities

Outline: *Up until now, we have assumed that the speed function F is defined not only on the zero level set, but on all level sets throughout the computational domain. In many applications, particularly those which link these interface techniques with underlying physics, this is not so. Consequently, we need an "extension velocity" which, starting with a given velocity prescribed at the interface, builds an appropriate velocity field everywhere in the computational domain. In this chapter, we explain the technique of Adalsteinsson and Sethian [6] which merges Fast Marching Methods and level set methods in order to build fast and appropriate extension velocities for the neighboring level sets.*[1]

11.1 The need for extension velocities

Let us begin by recalling the basic idea of Level Set Methods, namely that an interface propagating with speed F is embedded as the zero level set of the higher-dimensional level set function ϕ, yielding

$$\phi_t + F|\nabla\phi| = 0. \tag{11.1}$$

A somewhat hidden point is that the speed F is assumed to have been defined for <u>all</u> the level sets, not just the zero level set corresponding to the interface. Thus, not only is the interface embedded in a higher dimensional function, but the speed F of the interface is **itself** embedded in a higher-dimensional function. More accurately, we should write

$$\phi_t + F_{\text{ext}}|\nabla\phi| = 0, \tag{11.2}$$

---

[1] This material is taken from [6] with little change.

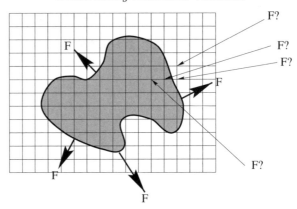

Fig. 11.1. Building extension velocities.

where F_{ext} is some velocity field which, at the zero level set, equals the given speed F. In other words,

$$F_{\text{ext}} = F \text{ at points } (x, y) \text{ where } \phi = 0$$

This new velocity field F_{ext} is known as the "extension velocity."

How much freedom does one have in this extension construction (see Figure 11.1)? In fact, there is considerable room to maneuver. The extension velocity F_{ext} should, in the limit as one approaches the zero level set, yield the speed F of the zero level set, i.e.,

$$\lim_{x \to a} F_{\text{ext}}(x) = F(a), \tag{11.3}$$

where a is a point on the front. Beyond these requirements, there is considerable freedom. We now provide a fast methodology for constructing a particular choice of extension velocity. The technique comes from exploiting the Fast Marching Method.[2]

There are three reasons for building our particular choice of extension velocity.

(i) **No natural speed function:**. In some physical problems, the velocity is given only at the front itself. For example, semiconductor manufacturing simulations of the etching and deposition process require determination of the visibility of the interface with respect to the etching/deposition beam (see [3, 4, 5]).

---

[2] The methodology may be used in either the Cartesian or triangulated settings. It is somewhat more straightforward in the triangulated case, and hence we will focus on explaining it in the Cartesian framework.

There is no natural velocity off the front, since it is unclear what is meant by the "visibility" of the other level sets. In this case, an extension velocity must be specifically constructed.

(ii) **Sub-grid resolution:**. As demonstrated in a later example, there are problems in which the speed of the interface changes very rapidly or discontinuously as the front moves through the domain. In such cases, the exact location of the interface determines the speed, and constructing a velocity from the position of the interface itself, rather than from the coarse grid velocities, is desirable.

(iii) **Maintaining a nice level set representation:**. Under some velocities, such as those which arise in fluid mechanics simulations [54, 262, 282], the level sets have a tendency to either bunch up or spread out, which is seen when ϕ becomes either very steep or flat. The extension velocity discussed here is designed so that an initial signed distance function is essentially maintained as the front moves; this will be demonstrated in Chapter 12. The reason to maintain the signed distance function is because, by keeping a uniform separation for the level sets around the front, calculation of variables such as curvature becomes more accurate. Our algorithm avoids all re-initialization, which can often perturb the front, and the problem of bunching or stretching is greatly ameliorated.

In the rest of this chapter, we explain the approach devised by Adalsteinsson and Sethian [6]. We refer the reader to that paper for a more detailed explanation and many examples.

11.2 Various approaches to extension velocities

The original level set calculations in [187] were concerned with interface problems with geometric propagation speeds, and hence an extension velocity was naturally built by using the geometry of each given level set. In more non-geometric/local applications, many different extension velocities have been employed. In many fluid simulations, one can choose to directly use the fluid velocity itself to act as F_{ext}. This is what was done by Rhee, Talbot, and Sethian [202] in a series of simulations of turbulent combustion. They built an extension velocity using an underlying elliptic partial differential equation coupled to a source term along the interface. This was also done in the two phase flow simulations of Chen,

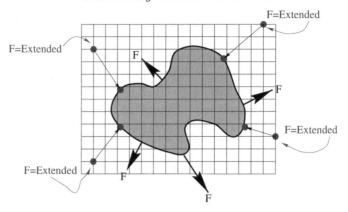

Fig. 11.2. Constructing extension velocities.

Hou, Merriman, and Osher [54] and Sussman, Smereka, and Osher [262]. In these detailed simulations, some bunching and flattening of the level set function occurs. This is repaired at every time step through a re-initialization process which rebuilds the signed distance function using an iterative process given in [262].

When there is no choice available for an extension velocity, Malladi, Sethian, and Vemuri in [169] suggested extrapolating the velocity from the front. Their idea is to simply extrapolate; standing at each grid point, the value of the speed function at the closest point on the front is used at the grid point. This technique requires location of the closest grid point (see Figure 11.2).

Another approach is to build a speed function from the front using some other, possibly less physical quantity. Sethian and Strain [243] developed a numerical simulation of dendritic solidification; in this model, the velocity at the interface depended on a jump condition across the interface and hence had no meaning for the other "non-physical" level sets. A boundary integral expression was developed for the velocity on the interface and evaluated both on and off the front to provide an extension velocity. The crystal growth study of Chen, Merriman, Osher, and Smereka [55] worked directly with the partial differential equations (rather than the conversion to a boundary integral), and built an extension velocity by solving an advection equation in each component, again coupled to a re-initialization procedure. We refer the interested reader to [55] for a collection of simulations performed using this approach.

The important point is that the velocity field F_{ext} used to move the level sets neighboring the zero level set need have nothing to do with the velocity suggested by the physics in the rest of the domain. It need only agree with the velocity F at the zero level set corresponding to the interface.

11.3 Equations for extension velocities

What are desirable properties of an extension velocity? First, it should match the given velocity on the front itself. Second, it is desirable if it moves the neighboring level sets in such a way that the signed distance function is preserved. Consider for a moment an initial signed distance function $\phi(x, t = 0)$, and suppose one builds an extension velocity which satisfies

$$\nabla F_{\text{ext}} \cdot \nabla \phi = 0. \qquad (11.4)$$

It is straightforward to show that under this velocity field, the level set function ϕ remains the signed distance function for all time, assuming that both F and ϕ are smooth. To see that this is so (see [282]), suppose that initially $|\nabla \phi(x, t = 0)| = 1$, and we move under the level set equation $\phi_t + F_{\text{ext}} |\nabla \phi| = 0$; then we note that

$$\frac{d|\nabla \phi|^2}{dt} = \frac{d}{dt}(\nabla \phi \cdot \nabla \phi) = 2\nabla \phi \cdot \frac{d}{dt}\nabla \phi$$

$$= -2\nabla \phi \cdot \nabla F_{\text{ext}} |\nabla \phi| - 2\nabla \phi \cdot \nabla |\nabla \phi| F_{\text{ext}}.$$

The first term on the right is zero because of the way the extension velocity is constructed; the second is zero because $|\nabla \phi(x, t = 0)| = 1$. Thus, one solution satisfies $|\nabla \phi| = 1$; this plus a uniqueness result for this differential equation shows that $|\nabla \phi| = 1$ for all time.

Thus, the strategy introduced in [6] is as follows. We shall use a two-tiered system. Given a level set function at time n, namely ϕ_{ij}^n, we first construct a signed distance function $\bar{\phi}_{ij}^n$ around the zero level set using the Fast Marching Method. Simultaneous with this construction, we construct the extension velocity F_{ext} satisfying Eqn. 11.4. We then use this velocity to update the level set function ϕ^n.

There are several important things to note about this approach:

- This construction finds an extension velocity which is then used to update the level set function. One can, of course, use a method of as high an order method as one chooses for the level set update. If one wants to perform this update restricted to a narrow band using the narrow band methodology of [2], one is free to do so. However, this methodology provides a way of doing so in all of space very rapidly, i.e., $O(N \log N)$, where N is the total number of points where one wants to build this extension velocity.

- In this approach, one can choose never to re-initialize the level set function as follows:

  (i) Consider a level set function ϕ^n at time step $n\Delta t = 0$.

  (ii) Build the extension velocity by simultaneously constructing a temporary signed distance function ϕ^{temp} and an extension velocity such that

  $$\nabla \phi^{\mathrm{temp}} \cdot \nabla F_{\mathrm{ext}} = 0,$$

  with ϕ^{temp} matching ϕ^n at their zero level sets, and F_{ext} matching the F given on the interface.

  (iii) Then advance the level set function ϕ^n under the computed extension velocity to produce a new ϕ^{n+1} by solving $\phi_t + F_{\mathrm{ext}}|\nabla \phi| = 0$.

This algorithm never re-initializes the evolving level set function, yet moves it under a velocity field that maintains the signed distance function. This avoids a large set of problems that have plagued some implementations of level set methods, namely that re-initialization steps can perturb the position of the front corresponding to the zero level set.

- In this approach, we explicitly find the zero level set corresponding to the interface in order to build the extension velocity. This may seem slightly "illegal": one of the appealing features of level set methods is that the front need not be explicitly constructed and that all of the methodology may be executed on the underlying grid. Here, we choose to explicitly build the front. However, we neither move nor update that representation. In cases of speed functions that depend on factors like visibility, this is completely natural. The central virtue of level set methods lies in the update of the level set function on a discrete mesh to embed the motion of the interface itself. This strategy and philosophy are maintained.

To summarize, given a front velocity F, this choice of extension velocity allows one to update an interface represented by an initial signed distance function in such a way that the signed distance function is maintained, and the front is never re-initialized. If one chooses to use the adaptive methodologies given in the narrow band approach, occasional rebuilding of the narrow band may be required, but this is performed only occasionally.

11.4 Building extension velocities

Recall that given a level set function ϕ^n, our goal is to build an extension velocity F_{ext} such that if $|\nabla\phi| = 1$, then updating under this extension velocity maintains this unit gradient. We solve the equation

$$\nabla\phi^{\text{temp}} \cdot \nabla F_{\text{ext}} = 0$$

so that ϕ^{temp} is the signed distance function which has the same zero level set as the level set function ϕ^n. We stress that we do not use this computed signed distance to re-initialize the level set function; it is used instead only in the construction of F_{ext}. We now explain this construction.

11.4.1 Constructing signed distances

Suppose we are given a level set function ϕ^n, where the superscript n indicates the time step in the usual notation, and suppose that this level set function does not correspond to the signed distance function. We can use the Fast Marching Method to compute the signed distance ϕ^{temp} by solving the Eikonal equation

$$|\nabla T| = 1$$

on either side of the interface, with the boundary condition that $T = 0$ on the zero level set of ϕ. The solution T will then be our temporary signed distance function ϕ^{temp}. The Fast Marching Method is run separately for grid points outside and inside the front (we note that whether a grid point is inside or outside is immediately apparent from the level set function ϕ^n).

The only difficulty is in the initialization stage of the Fast Marching Method, that is, the computation of the approximate distances of the set of *Trial* points in order to begin the Fast Marching Method.[3] We now show how to find the initial set of *Trial* values for grid points outside of a two dimensional front; points inside the front and points close to a three-dimensional surface are handled similarly.

Begin by initially tagging as *Trial* those grid points where one of the

---

[3] See the section on Fast Marching Methods for the meaning of the notation *Trial*, etc.

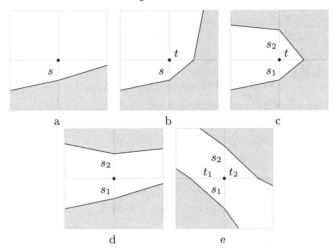

Fig. 11.3. All cases for the neighborhood of a point.

neighbors lies inside the front. We must assign values at these points
to approximate the distances to the front. While this can be computed
exactly for a smooth front, a faster method can be designed which uses
only the intersection of the front with the grid lines. This is particularly
useful when the front is given as the zero level set of a function defined
at the grid points, and a smooth representation is not available.

Up to rotation there are five possible cases that need to be considered,
as are shown in Figure 11.3.

- In Figure 11.3a, only one of the neighboring points is on the other
  side of the front. Here we define the value to be the distance s to the
  intersection point on the line connecting the two grid points. This
  value is larger than the real distance to the front, but most likely
  the value at the grid point on the other side is the distance to the
  same point, so that the zero level set will not have moved after the
  re-initialization.
- In Figure 11.3b, two of the neighbors are on the other side of the
  front. In this case the value is defined as the exact distance to the
  line segment between the two intersection points. If s and t are the
  distances to the intersection points, the exact distance d satisfies

$$\left(\frac{d}{s}\right)^2 + \left(\frac{d}{t}\right)^2 = 1.$$

The left-hand-side is an upwind approximation to the gradient of the distance function, since the distance is zero at the intersection points. This suggests what the solution should be for the remaining three cases, and how it should be computed in three dimensions.

- In Figure 11.3c, the distance is the positive solution to

$$\left(\frac{d}{\min(s_1, s_2)}\right)^2 + \left(\frac{d}{t}\right)^2 = 1.$$

- In Figure 11.3d, the distance is

$$d = \min(s_1, s_2).$$

- In Figure 11.3e, the distance is the positive solution to

$$\left(\frac{d}{\min(s_1, s_2)}\right)^2 + \left(\frac{d}{\min(t_1, t_2)}\right)^2 = 1.$$

11.4.2 Constructing the velocity extension F_{ext}

We now extend a speed function given along an interface to grid points around the front. This extension should extend the speed in a continuous manner, and avoid, if possible, the introduction of any discontinuities in the speed close to the front. Recall that we want to construct a speed function F_{ext} that satisfies the equation

$$\nabla F_{\text{ext}} \cdot \nabla \phi^{\text{temp}} = 0.$$

The idea is to march outward using the Fast Marching Method, simultaneously attaching to each grid point both the distance from the front and the extended speed value. We first compute the signed distance ϕ^{temp} to the front using the Fast Marching Method as described in the previous section. As the Fast Marching Method constructs the signed distance at each grid point, we simultaneously update the speed value F_{ext} according to Eqn. 11.4.2. In the gradient stencil, we use only neighboring points close to the front to maintain the upwind ordering of the point construction.

In more detail, and similar to the construction of signed distances, we first need to find the speed values for the initial set of *Trial* points in order to begin the technique, and then update the extension value when the distance value is updated according to the above equation.

One technique for building extension velocities near the front is to copy the speed of the closest grid point. Instead, we take a weighted average of the speed values at the points which are used in computing the distance, where the weight is proportional to one over the square of the distance; this is equivalent to solving the equation $\nabla F_{\text{ext}} \cdot \nabla \phi^{\text{temp}} = 0$.

As an example, consider the cases in Figure 11.3. For simplicity, assume that we are computing the extension value for the point (i, j) in the center.

- For Figure 11.3a, the extension speed is $f = f(i, j - s)$.
- For Figure 11.3b, the gradient is given by

$$\left(-\frac{d}{t}, \frac{d}{s} \right).$$

The equation $\nabla F_{\text{ext}} \cdot \nabla \phi^{\text{temp}} = 0$ is then given by

$$
\begin{aligned}
0 &= \left(-\frac{f - f(i + t, j)}{t}, \frac{f - f(i, j - s)}{s} \right) \cdot \left(-\frac{d}{t}, \frac{d}{s} \right) \\
&= d \left[\frac{f - f(i + t, j)}{t^2} + \frac{f - f(i, j - s)}{s^2} \right],
\end{aligned}
$$

in which case

$$f = \frac{\frac{1}{t^2} f(i + t, j) + \frac{1}{s^2} f(i, j - s)}{\frac{1}{t^2} + \frac{1}{s^2}}.$$

This equation indicates the solution for the remaining cases and for the three-dimensional case. Our expression assumes that the speed of the interface is given at the intersection points of the interface with the grid lines. If the speed is given at other points, one can either use interpolation to get the speed values or modify the previous algorithm.

- For Figure 11.3c, the equation is

$$f = \frac{\frac{1}{t^2} f(i + t, j) + \frac{1}{s^2} f(i, j + s)}{\frac{1}{t^2} + \frac{1}{s^2}},$$

where $s = s_1$ if $|s_1| < |s_2|$; otherwise $s = s_2$.

- For Figure 11.3d, the equation is

$$f = f(i, j + s),$$

where s is chosen as in the term before.

- For Figure 11.3e, the equation is

$$f = \frac{\frac{1}{t^2} f(i + t, j) + \frac{1}{s^2} f(i, j + s)}{\frac{1}{t^2} + \frac{1}{s^2}},$$

where s and t are chosen between the entry from $\{s_1, s_2\}$ and $\{t_1, t_2\}$ which are smaller in absolute value.

Once values for both the signed distance and the extension function are established at *Trial* points, we need only update extension values. As the distance value is updated using the Fast Marching Method, a new extension value is chosen such that $\nabla F_{\text{ext}} \cdot \nabla \phi^{\text{temp}} = 0$, where the gradient of F_{ext} and ϕ^{temp} are calculated using the points that contributed in the update of ϕ. If no points from a grid direction are used, the corresponding component of the gradient is zero.

As an example, consider the case shown in Figure 11.3b. Here the new distance value at (i, j) is found by solving Eqn. 6.31. Assuming that $(i+1, j)$ and $(i, j-1)$ are the points that are used in updating the distance, if v is the new extension value, it then has to satisfy an upwind version of Eqn. 8.2, namely

$$\left(\frac{\phi^{\text{temp}}_{i+1,j} - \phi^{\text{temp}}_{i,j}}{h}, \frac{\phi^{\text{temp}}_{i,j} - \phi^{\text{temp}}_{i,j-1}}{h} \right) \cdot \left(\frac{F_{i+1,j} - v}{h}, \frac{v - F_{i,j-1}}{h} \right) = 0.$$

Since $(i+1, j)$ and $(i, j-1)$ are known, F is defined at those points, and this equation can be solved with respect to v to produce

$$v = \frac{F_{i+1,j}(\phi^{\text{temp}}_{i,j} - \phi^{\text{temp}}_{i+1,j}) + F_{i,j-1}(\phi^{\text{temp}}_{i,j} - \phi^{\text{temp}}_{i,j-1})}{(\phi^{\text{temp}}_{i,j} - \phi^{\text{temp}}_{i+1,j}) + (\phi^{\text{temp}}_{i,j} - \phi^{\text{temp}}_{i,j-1})}.$$

This means that, at the end, $\nabla F_{\text{ext}} \cdot \nabla \phi^{\text{temp}} = 0$, is satisfied for all points on the grid, except the points along the front itself. At those points, the previous construction will satisfy the equation when the gradient approximation is computed using points on the front, as mentioned earlier. For complete details, see [6].

11.5 A quick demonstration

As a quick demonstration of extension velocities (we shall return to this topic many times in the remainder of the book), consider an initial circle expanding with speed

$$F(x, y) = ay + b,$$

where a and b are chosen so that $F(x, 2) = 1.0$ and $F(x, 10) = 10.0$. Thus, the motion elongates the circle as it propagates outward. In Figure 11.4, we show the results of this calculation. On the left, we show the initial front, together with the front at time $T = 0.633$ and the other level sets. The initial front is shown as a light dashed line, while the front at a later time is shown as a dark dashed curve. This calculation is performed without extending the front velocity to the neighboring

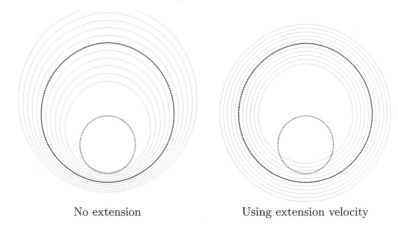

No extension   Using extension velocity

Fig. 11.4. Motion under elongation speed function: the small dark circle in each figure is the initial front. The larger dark circle and the neighboring level curves are computed at time $t = 0.633$. There is no re-initialization nor rebuilding of the evolving level set function in either calculation.

curves. Instead, the speed F_{ext} is taken directly from the analytic function given earlier. On the right, the same calculation is repeated by extending the velocity off of the front using our algorithm. The figure on the left shows stretching of the neighboring level sets, while the one calculated using our extension velocity algorithm maintains nicely the signed distance function, as seen by the evenly spaced level set function lines.

11.6 Re-initialization

"Re-initialization" refers to the idea of stopping a level set calculation at some point in time and rebuilding a level set function corresponding to the signed distance function. There are several reasons one might choose to do this.

- First, as part of the Narrow Band Level Set Method, if the front approaches the edge of the computational domain, one typically wants to stop the calculation and rebuild a new Narrow Band which evenly straddles the zero level set corresponding to the front. In the case of a very small narrow band, as was utilized in [54, 262], this re-initialization must be done at virtually every time step.
- Second, suppose one chooses to use a speed function defined on the neighboring level sets which does not preserve the signed distance

function as the level set function evolves. For example, in simulations that directly use a fluid velocity calculated on the entire computational domain (see [202, 54, 262]), the level sets that neighbor the zero level set move with velocities that considerably distort and stretch the field ϕ. In this case, one must re-initialize the level set function very often in order to rebuild this signed distance function.

There are several ways to accomplish this re-initialization. In Chopp's work on minimal surfaces [59], he needed to reattach level sets to a bounding wire frame. His technique for re-initializing stopped the calculation and simply recomputed the signed distance function by standing at each grid point in a narrow band and finding the distance to the front. This sounds expensive; however, judicious use of quadtree and octree structures, common in particle methods and adaptive mesh refinement, make this a viable technique.

Another approach is to use Fast Marching Methods, as described earlier in this chapter. One runs the Fast Marching Method first forward, and then backward, in both cases solving the Eikonal equation:

$$|\nabla T| = 1.$$

The computed signed distance function ϕ^{temp} is simply substituted for the current level set function, providing a very fast way of re-initializing beyond one or two cells away from the front.

A different technique was introduced by Sussman, Smereka, and Osher [262], based on an observation of J.-M. Morel. Its virtue is that the level set function is re-initialized without explicitly finding the zero level set. Consider the partial differential equation

$$\phi_t = \text{sign}(\phi)(1 - |\nabla \phi|), \tag{11.5}$$

where $\text{sign}(\phi)$ gives the sign of ϕ. Given any initial data for ϕ, solving this equation to steady state provides a new value for ϕ with the property that $|\nabla \phi| = 1$, since convergence occurs when the right-hand side is zero. The sign function controls the flow of information in Eqn. 11.5. If ϕ is negative, information flows one way, and if ϕ is positive, then information flows the other way. The net effect is to "straighten out" the level sets on either side of the zero level set and produce a ϕ function with $|\nabla \phi| = 1$ corresponding to the signed distance function. Thus, the strategy is to stop the level set calculation periodically and solve Eqn. 11.5 until convergence. If done often enough, the initial guess is often close to the signed-distance function and few iterations are required. This technique was used extensively in [262, 54, 55, 176, 282] as part of some fluid and materials sciences calculations.

One potential disadvantage of this iterative re-initialization scheme is the relative crudeness of the switch function based on checking the sign of the level set equation. Considerable motion of the zero level set can occur during the re-initialization, since the sign function does not do an accurate job of using information about the exact location of the front. As pointed out by Sussman and Fatemi [259], this approach has the problem that the "larger the number of iterations, the more the zero level set would stray from the expected position." In other words, the re-initialization algorithm causes the front to move. This can manifest itself in an unnecessary loss of mass in fluid flow calculation. A fix for this problem was introduced by Sussman and Fatemi in the same paper ([259]); by solving an associated variational problem during the iteration, the total amount of mass can be conserved, and the position of the zero level set remains almost unchanged during the iterations. The variational construction makes sure that the iteration algorithm is not allowed to change the total amount of mass in any given cell by enforcing a constraint. Sussman and Fatemi have shown some nice results using their algorithm, and their approach is a good idea if one chooses to re-initialize every time step using an iterative approach.

Nonetheless, our suggested approach is to avoid re-initialization as much as possible. Problems with loss of mass and movement of the zero level set can be avoided if one uses the extension velocity methodology presented in this chapter. Since the extension methodology produces a velocity field which essentially maintains the signed distance function, and because this approach never replaces the evolving level set function, it provides an accurate, straightforward, and highly efficient way of advancing fronts under complex velocity fields. By doing so, re-initialization is sparingly reserved only for occasional rebuilding of narrow bands.

Tests of Basic Methods

Outline: *We perform some basic tests of the numerical accuracy of Level Set Methods and Fast Marching Methods. The goal is to examine the accuracy of first and second order schemes for the level set equation, to measure grid effects and conservation of mass, to analyze the accuracy of triangulated Level Set Methods based on both positivity schemes and Petrov-Galerkin methods, to examine first and second order Fast Marching Methods, and to test various aspects of the extension velocity methodology.*

12.1 The basic Cartesian Level Set Method

We begin with tests of the basic first and second order Level Set Method introduced in [187]. In order to do so, we must take into account three separate sources of error:

(i) **Initialization error:** Error associated with the building the initial signed distance function on a discrete mesh from a smooth initial curve.

(ii) **Update error:** Error associated with updating the time-dependent level set equation.

(iii) **Measurement error:** Error associated with extracting the position of the front $\phi = 0$ from the level set function; this may involve the use of a contour plotter or other diagnostics.

Deciding how to remove the first and third sources of error in this list requires some thought. As an example, consider the very simple case of a circle expanding with unit speed, which will serve as one of our test examples. The first issue is to build a discrete signed distance function

141

Scheme/Grid	21x21	41x41	81x81
First Order	.0360907	.0187256E	.00226235
Second Order	.012602	.0031896	.0008339

Error = Computed crossing time at (1,1) - exact crossing time

Fig. 12.1. Grid orientation effects.

$\phi(x, y, t = 0)$ on a uniform grid of mesh size h centered at the origin. One can think of at least three ways to build this signed distance function:

(i) At each grid point i, j, compute the distance d_{ij} to the origin, and let the initial signed distance be defined by $\phi_{ij}^{n=0} = d_{ij} - 1$.

(ii) Standing at each grid point i, j, numerically compute the signed distance r_{ij} to the initial unit circle, and let $\phi_{ij}^{n=0} = r_i$.

(iii) Construct values for the initial signed distance function $\phi_{ij}^{n=0}$ so that the <u>area</u> enclosed by the level curve $\phi = C$, when found using some chosen form of numerical quadrature, will exactly equal the correct enclosed area; one can think of many variants of this idea.

Note the differences in these ideas. The first initialization utilizes the exact analytic initial data. The second contains the numerical error associated with finding the distance to the initial curve, assuming that this is not done analytically. The third focuses on the error associated with the eventual diagnostic used to analyze the result.

We shall use the first technique; that is, to construct grid values with regard to the exact distance to the origin. Our interest therefore will be in the decay in accuracy as the solution evolves, not in the initial error. For the purposes of our tests, we shall also ignore measurement error.

First, we test the accuracy of the first and second order schemes. We begin with a circle of radius 0.5 as the initial datum, and allow it to expand with speed $F = 1$. The computation of the position of the front along either the x or y axis will be almost exact, since the problem defaults to one-dimensional quadrature. Hence, to both measure the error and analyze grid effects, in Figure 12.1 we compute the crossing time when the front (that is, the zero level set) crosses the point $(1,1)$; obviously this should happen at $T = \sqrt{2} - 0.5$. We choose a time step of 0.9 times the maximum CFL number allowed. As the error vanishes under grid refinement, the front becomes circular and shows little effect of the grid orientation. The distortion is considerably less for the second order scheme, and, as the mesh is refined, the distortion of the circle is lessened.

Scheme/Grid	11x11	21x21	41x41	81x81
First order	8.639%	4.038%	1.987%	.9905%
Second order	2.845%	.64671%	.28064%	.02904%

Fig. 12.2. Percentage change in total area at $T = 1$ under translation.

Next, we check conservation of mass. We begin by noting that there has been considerable discussion in the literature about the conservation of mass properties of level set methods. Of all front tracking techniques, volume-of-fluid schemes are the most conservative methods, because they are explicitly built to conserve mass. Any fluid that is removed from one cell must be placed in another cell, and hence the volume fractions are conserved. However, this does not ensure that the mass goes to the right place; it only guarantees that the total amount is conserved.

There are two separate sources of error in mass conservation for level set methods. The first is the numerical error associated with the scheme used to advance the partial differential equation. The second is the choice of velocity field used in this update. From now on, we will always use the extension velocity approach described in the previous chapter. We begin with the simple problem of a circle advecting with velocity $(1, 0)$. Thus, the solution is just a circle translating to the right, and the shape should remain unchanged; again, we use our extension velocity approximation to build speed functions for the neighboring level sets. To test the level set method, we will view this as a front propagating in its normal direction with speed

$$F = (1, 0) \cdot \left(\frac{\nabla \phi}{|\nabla \phi|}\right),$$

and thus we solve

$$\phi_t + (1, 0) \cdot \left(\frac{\nabla \phi}{|\nabla \phi|}\right)|\nabla \phi| = 0.$$

In Figure 12.2, the results of this experiment are shown. The circle is moved until time $T = 1.0$, and then the percentage change in the area is measured. As can well be seen, the error decreases as the mesh is resolved, and higher order schemes yield more favorable results. We note that there is some arbitrariness in the choice of time step. For the first order method, we choose 0.9 times the maximum allowed by the Courant condition; for the second order method, we choose a time step which provides the best value for the mass conservation. Of course, since

Scheme/Grid	21x21	41x41	81x81	161x161
First order	51.88%	37.981%	16.77%	6.14%
Second order	18.00%	2.276%	.5999%	.09758%

Percentage Change in Total Area at $T = 1$ under solid body rotation.

Fig. 12.3. Conservation of mass.

the flow is essentially translation to the right, the quality of the results must be viewed with some caution.

A more difficult test comes from considering the solid body rotation of a circle of radius one centered at $(-1, 0)$. We consider a half rotational cycle of the circle so that it returns to a point opposite its starting point. We use a velocity field $(u, v) = (-y, x)$, again projected onto the normal to the front, so that we are solving

$$\phi_t + F|\nabla \phi| = 0,$$

with $F = (-y, x) \cdot \frac{\nabla \phi}{|\nabla \phi|}$. In Figure 12.4 we show the motion of this circle for various grid refinements under both first and second order methods. We note that for crude enough meshes, considerable diffusion takes place, and the front motion is limited. As the meshes are refined, we approach almost complete conservation of mass, as can be seen in the table given in Figure 12.3. This table shows the percentage change in the enclosed area. We note that for both very crude and very fine meshes, the error associated with measuring the enclosed area contributes to variations in the measured values. Even better results can be achieved by artfully picking the optimal time step.

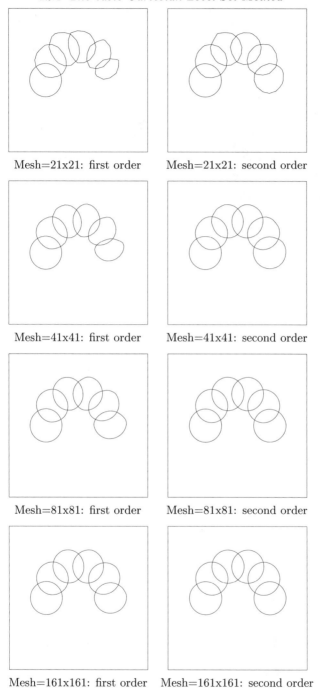

Mesh=21x21: first order   Mesh=21x21: second order

Mesh=41x41: first order   Mesh=41x41: second order

Mesh=81x81: first order   Mesh=81x81: second order

Mesh=161x161: first order   Mesh=161x161: second order

Fig. 12.4. Solid body rotation of unit circle under mesh refinement.

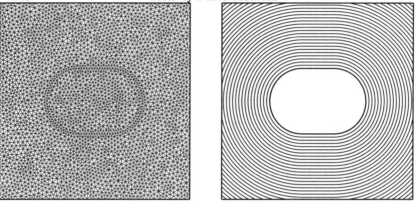

Fig. 12.5. Smooth level set solutions: coarsest mesh (left) and crossing time solution (right).

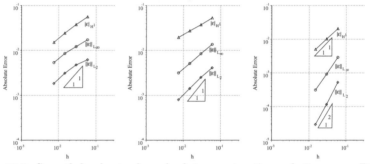

Fig. 12.6. Smooth level set solns: absolute crossing time solution error. Positive coefficient scheme (left), explicit Petrov-Galerkin (middle), and implicit Petrov-Galerkin (right).

12.2 Triangulated Level Set Methods for H–J equations.

Next, we study some aspects of the error associated with the triangulated level set and Hamilton-Jacobi algorithms. Here, we merely summarize some of the numerical tests given in [24] and refer the interested reader there for more details.

12.2.1 Smooth solutions

As a first example, the level set equation is solved starting from smooth initial solution data corresponding to an oval. Exterior to the oval, the solution remains smooth for the entire calculation.

Figure 12.5 shows the mesh and crossing time solution computed using the positive coefficient scheme. We do not compute the crossing time so-

Fig. 12.7. Non-smooth level set solutions: coarsest mesh (left) and crossing time solution (right).

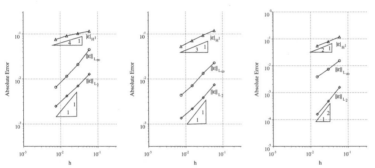

Fig. 12.8. Non-smooth level set solutions: absolute crossing time solution error. Positive coefficient scheme (left), explicit Petrov–Galerkin (middle), and implicit Petrov-Galerkin (right).

lution interior to the specified oval curve. Graphs of the absolute crossing time error using the explicit positive coefficient, explicit Petrov-Galerkin, and implicit Petrov-Galerkin schemes are shown in Figure 12.6. The two explicit schemes yield slightly less than first order accuracy. In contrast, the implicit Petrov–Galerkin scheme yields full second order accuracy in L_2 and L_∞ norms and first order accuracy in the H^1 semi-norm. These are optimal convergence rates in these norms for linear elements.

12.2.2 *Non-smooth solutions*

As a second example, the level set equation is solved starting from smooth initial solution data corresponding to a star-shaped curve which remains only Lipschitz smooth during time evolution.

Figure 12.7 shows the mesh and crossing time solution computed using the positive coefficient scheme. Note the non-oscillatory resolution of the slope-discontinuous corner. Graphs for absolute error in the crossing time solution are shown in Figure 12.8. The two explicit schemes again show slightly less than first order accuracy in L_2 and L_∞ with further degradation in the H^1 semi-norm. The implicit Petrov-Galerkin with discontinuity capturing term parameter $\sigma_{dc} = .1$ retains second order accuracy in L_2, first order accuracy in L_1, and one-half order accuracy in the H^1 semi-norm.

12.2.3 *Mesh adaptation and curvature flow*

Next, we show examples of mesh adaptation and curvature flow. Figure 12.9 shows three different uses of mesh adaptation for evolving front problems. Figure 12.9a shows conformal adaptation around the front itself; Figure 12.9b shows Steiner adaptation around regions of high curvature in the evolving front; Figure 12.9c shows interface conforming Steiner adaptation around regions of high curvature in the evolving solution.

Next, we consider a speed function of the form $F = -\kappa$, where κ is the two-dimensional curvature. This will be discussed in more detail in the applications section. For now, we simply note that such a flow has some sort of regularizing effect. In Figure 12.10, we show a mesh adapted around the zero level set in the top row, and the resulting interface in the row below.

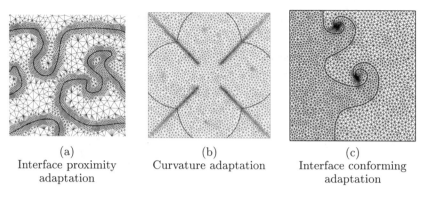

<table>
<tr><td>(a)</td><td>(b)</td><td>(c)</td></tr>
</table>

(a)	(b)	(c)
Interface proximity	Curvature adaptation	Interface conforming
adaptation		adaptation

Fig. 12.9. Various mesh adaptations for evolving interface problems.

Fig. 12.10. Mesh refinement and curvature flow.

12.3 Accuracy of Fast Marching Methods

Next, we examine the accuracy of the Fast Marching Method. The "second order" scheme is the one which uses second order one-sided operators whenever possible. Figure 12.11 shows an equi-distance set of points from the origin, computed using (i) the exact distance, (ii) the first order scheme and (iii) the "second order" scheme on an extremely crude 6x6 grid. As can be seen, the "second order" method is considerably better. Figure 12.11 also gives the error associated with computing the distance function from a single point located at the origin.

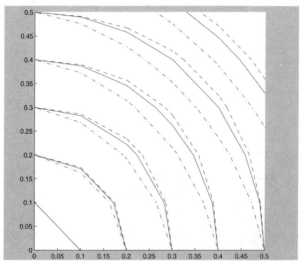

Equidistant points from origin: crude 6x6 Grid
Solid = exact, Short dashed/dotted = 1st order, Long dashed = 2nd order.

Grid	L_2 Error 1st Order	L_2 Error 2nd Order	L_∞ Error 1st Order	L_∞ Error 2nd Order
21^3	0.019790	0.007203	0.034042	0.012921
51^3	0.009409	0.002410	0.017063	0.004325
101^3	0.004610	0.000461	0.008570	0.000734
151^3	0.002289	0.000071	0.004325	0.000205

Fig. 12.11. First and second order computations of distance from origin.

Next, we repeat this test, but use the Fast Marching Method to compute the distance from two points, one located at $(0.5, 0.25)$ and the other at $(0.5, 0.75)$. Here, the solution is non-differentiable, and the viscosity solution is chosen. Once again, the "second order" method is considerably more accurate. Figure 12.12 shows a calculation on a rough 21x21 grid; the table in Figure 12.12 below shows the error under mesh refinement.

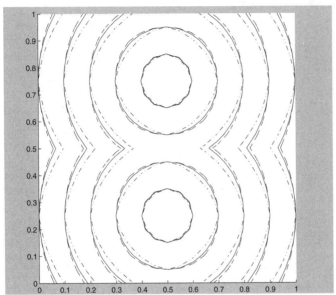

Equidistant points from two points: crude 21x21 Grid
Solid = exact, Short dashed/dotted = 1st order, Long dashed = 2nd order.

Grid	L_2 Error 1st Order	L_2 Error 2nd Order	L_∞ Error 1st Order	L_∞ Error 2nd Order
21^3	0.0096	0.00500	0.0185	0.01047
41^3	0.0051	0.00119	0.0101	0.00230
81^3	0.0028	0.00028	0.0056	0.00055
161^3	0.0014	0.00006	0.0028	0.00014

Fig. 12.12. First and "second order" computations of distance from two points.

Finally, we repeat this test using a grid rotation and compute the distance from two points, one located at $(0.25, 0.75)$ and the other at $(0.75, 0.25)$. The L_2 error is still second-order convergent for the second order scheme, while the L_∞ error is first order. This is to be expected; first order error occurs along the shock lines, which, due to the grid alignment, occurs between grid points. Figure 12.13 shows a calculation on a rough 21x21 grid; the table in Figure 12.13 below shows the error under mesh refinement.

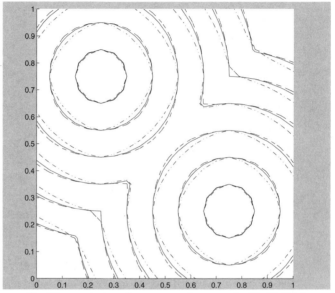

Equidistant points from two points: crude 21x21 Grid
Solid = exact, Short dashed/dotted = 1st order, Long dashed = 2nd order.

Grid	L_2 Error 1st Order	L_2 Error 2nd Order	L_∞ Error 1st Order	L_∞ Error 2nd Order
21^3	.01410	.005329	.02534	.013800
41^3	.00646	.001832	.01228	.005449
81^3	.00311	.000409	.00607	.002538
161^3	.0015	.000113	.00303	.001346

Fig. 12.13. First and "second order" computations of distance from two points.

12.4 Tests of extension velocity methodology

Next, we perform numerical tests of the extension velocity construction, designed to examine the following:

- How much time does it take to construct the extension velocities?
- How well do extension velocities maintain the signed distance function?
- What is the advantage of using this technique in problems with subgrid effects?

By way of description, we denote by "extension method" the Fast Marching Method extension algorithm given previously, while defining the "no extension" method to be the one which uses a velocity defined and available at each particular level set. Also, we shall distinguish between the Narrow Band Method which limits computation to the narrow band, and the full matrix method, which operates on the entire computational domain. The tests in this section are taken without change from [6], and we refer the interested reader there for further details.

12.4.1 Timings

Tube width	100x100 grid		300x300 grid	
	# Points in domain	Time	# Points in domain	Time
7 cells	1500	8.8ms	4740	52ms
11 cells	2340	11.7ms	7660	66ms
15 cells	3124	14.8ms	10604	80ms
21 cells	4144	18.5ms	14780	100ms
Full domain	10000	39ms	90000	475ms

Fig. 12.14. Timings table for elongation flow.

We return to the previous example of the motion of an elongated circle motion and calculate the amount of time required to build the extension velocities. Figure 12.14 provides detailed timing measurements[1] and shows how the time required to build our extension velocity depends on the number of grid points where we want values. As the grid is refined, there are more points on the front, and hence the extension takes longer. We show results for various widths of the narrow band, and then finally for the "full" method which constructs extension velocities at all grid points in the full computational domain. As expected, the timings go up

---

[1] 350MHz - 604e - PowerPC processor

Grid Size	Narrow band		Full method	
	Extension	No extension	Extension	No extension
60	less than 1 sec	less than 1 sec	3s	less than 1 sec
120	5s	2s	22s	12s
240	30	9	180	96
600	450	94	3500	1600

Fig. 12.15. Simple advection with velocity $(2, 3)$.

as the number of grid points increases; the increase is essentially linear, since the $\log N$ term is almost one for these values.

The next question concerns the time required to construct the extension velocity take versus simply using the given value. Figure 12.15 compares the execution speeds[2] for the full method vs. extension approach for a narrow band of width 6 cells on either side of the interface.

These results show that the execution speed for the extension problem is roughly three to four times slower than the direct method. It has worse scaling, which is to be expected as the cost of extending the speed to M grid points is $O(M \log M)$. However, we note an important fact. It costs almost nothing to evaluate the speed function in this example. In problems where calculation of the speed function is *itself* time-consuming (for example, in the evaluation of a boundary integral to determine the speed function), minimizing the number of points at which this must be performed leads to significant speedup. As soon as the cost of computing the speed at the grid points is more than $O(\log M)$, the extension technique becomes competitive, and in fact can be faster than the direct method.

12.4.2 *Maintaining the signed distance function*

Next, we verify that our algorithm constructs an extension velocity which maintains the signed distance function, assuming that we start with an initialization which itself has the property that $|\nabla \phi(x, y, t = 0)| = 1$. We consider two separate problems, each designed to cause significant shearing, stretching, and compression of the neighboring level set functions.

First, we consider a speed function of the form

$$F = 2. * (R - (3 + 2t)) \sin(4\theta) + 2.$$

[2] Sun Sparc II.

Initial front   Level sets using given F   Level sets using extension velocity
   (a)   (b)   (c)

Fig. 12.16. Effect of no-extension and extension velocities on neighboring level sets: there is no re-initialization of the level set function at any time.

where $R = (x^2 + y^2)^{1/2}$, θ is the angle between the vector (x, y) and the positive x axis, and t is the time. Here, the initial circle is centered at the origin with radius 3. The calculation is performed on a 101×101 grid. We show (i) the initial values (Figure 12.16a), (ii) the position of the neighboring level sets at time $t = 0.8$ using the given velocity field ("no extension") (Figure 12.16b), and (iii) the results using our extension velocity ("extension") (Figure 12.16c). The use of the given velocity causes the level sets neighboring the zero level set (shown as a dark line) to wildly diverge, while using the extension velocity causes the level sets to remain equally spaced and conforming with the zero level set. One can imagine the difficulty involved in computing quantities like curvature from the neighboring level sets in the no-extension case. The exact solution to this problem (as can be seen from the velocity field) is a circle with radius given at time t by $(3 + 2t)$, which matches our computed solution.

We now repeat this calculation for another example and provide a quantitative measurement for the error involved. This time, we use a velocity field of the form

$$F = ((R - 3)^2 + 1) \cdot (2 + \sin(4\theta)).$$

The calculation is performed on a 401×401 grid. We start with the same initial data as given in Figure 12.17a. This velocity field produces significant distortions in the front. In Figure 12.17, we show the front at time $t = 0.45$ and $t = 0.90$ both without the extension and with our extension velocity. It is clear that the extension velocity preserves the

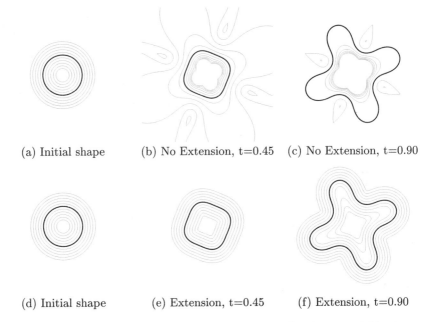

(a) Initial shape   (b) No Extension, t=0.45   (c) No Extension, t=0.90

(d) Initial shape   (e) Extension, t=0.45   (f) Extension, t=0.90

Fig. 12.17. Effects of no-extension and extension velocities on neighboring level sets (there is no re-initialization of the level set function at any time.)

initial property of the signed distance function, namely $|\nabla \phi| = 1$, as can be seen by the equal spacings of the neighboring level sets.

Once again, the equal spacing of the contour lines show that in the extension case, the signed distance function has been maintained. We note that the two computations of the non-zero level set contours are in fact supposed to be different; however, the zero level sets coincide.

Next, we analyze the error as a function of mesh size in the computed solution. We do this as follows. We assume that the finest calculation on an 800×800 grid of the extension velocity is the correct reference solution (for both first and second order methods) and compute the signed distance function from that zero level set at time $t = 0.90$. We do this as follows. We take the points on the reference path and for each point calculate the minimum distance to the path we want to compare it to. This gives an array of distances which are then used for the norm evaluations for the L_∞, L_1, and L_2 error. The calculation is made on a physical box of size 20×20. The results are shown in Figure 12.18 for a first and second order spatial method. As expected, the error in

First Order						
	No extension			Extension		
Grid	L_∞	L_1	L_2	L_∞	L_1	L_2
50^2	.00818	.00356	.00418	.00732	.00329	.00402
100^2	.00434	.00185	.00219	.00327	.00136	.00171
200^2	0.00211	.00090	.00111	.00137	.00056	.00071
400^2	.0011	.00047	.00059	.00045	.00018	.00023

Second Order						
	No extension			Extension		
Grid	L_∞	L_1	L_2	L_∞	L_1	L_2
50^2	.00365	.00138	.00163	.00279	.00107	.00138
100^2	.00103	.00042	.00049	.00075	.00027	.00036
200^2	.00039	.00016	.00019	.00019	.00006	.00009
400^2	.00025	.00008	.00011	.00004	0.00001	.00002

Fig. 12.18. Error analysis for no-extension and extension calculations.

the computed solution for the interface using the extension velocity is considerably better than that computed using the no-extension solution.

12.4.3 Capturing sub-grid effects

Finally, we show how this approach can help capture sub-grid effects by using the speed extension rather than the natural speed at the grid points.

As a first test, consider the problem of a front propagating with a speed F and subject to the constraint that the evolving interface cannot enter into a region Ω in the domain. This region Ω is referred to as a "mask," since it inhibits all motion. There are several solutions to this problem, depending on the degree of accuracy required.

The simplest solution is to set the speed function F equal to zero for all grid points inside Ω. The location of all points inside Ω can be determined before any calculation is carried out. This technique ensures that the front will stop within one grid cell of the mask. Figure 12.19 shows a plane front propagating with speed $F = 1$ in the upward direction, with a rectangular block in the center of the domain serving as a mask. In Figure 12.19(a), the speed function is reset to zero inside the mask region. As the front propagates upward it is stopped in the vicinity of the mask and forced to bend around it.

The calculations in Figure 12.19(a) are performed on a very crude 13×13 mesh in order to accentuate a problem with this approach; the

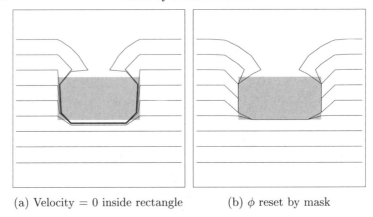

(a) Velocity = 0 inside rectangle   (b) ϕ reset by mask

Fig. 12.19. Front propagating upwards around masking block: 13×13 grid.

front is only guaranteed to stop within one grid cell of the obstacle itself. This is because the level set method constructs an interpolated speed between grid points. By setting the speed function to zero on and in the mask, the front slows down before it actually reaches the mask. Note that since this means one grid cell *normal* to the mask's boundary, a considerable amount of error can result.

A different fix, which eliminates some of this problem, comes from an alternate view; see Sethian [235]. Given a mask area Ω, construct the signed-distance function ϕ^{Ω} by taking the positive distance if inside Ω and the negative distance if outside (note that this is sign choice is the opposite of the one we typically use). Then we limit motion into the masked region, not by modifying the speed function, but instead by resetting the evolving level set function. Let ϕ^* be the value produced by advancing the level set ϕ^n one time step. Then let

$$\phi^{n+1} = \max(\phi^*, \phi^{\Omega}). \qquad (12.1)$$

This resets the level set function so that penetration is not possible; of course, this is accurate only to the order of the grid. Results using this scheme are shown in Figure 12.19(b). Again, we have used a very coarse grid to accentuate the differences. For the opposite problem, in which a region Ω acts as a source, the solution is equally straightforward and is given by $\phi^{n+1} = \min(\phi^*, -\phi^{\Omega})$; this is the technique used in [202].

While straightforward, none of these approaches is satisfactory; the evolving front still enters the mask to a considerable degree. Alternatively, we can use our extension velocity methodology and we extend from the value of the speed function F on the front itself, it picks up the correct velocity with subgrid accuracy and does not so closely depend on the computational grid used to update the level set function ϕ. As an illustration, consider a problem in which there is a dramatic range of speed values within a narrow spatial range. We take a domain in which the speed in the normal direction is 1, except in a thin annulus of width two grid cells where the speed is 100. As a practical motivation, one might consider an etching problem in which a very thin band of easily etched materials lies embedded within a material with much slower etch rate. We start with an initial front that intersects the circle, and we advect the front according to the given speed function. We note the difference between the extension and the direct approach:

- **Extension**: The speed at any point on the front can be calculated precisely by checking if it is inside this annulus or not. This speed is then extended onto the surrounding grid points. This leads to a more accurate detection of the front location and the accompanying etch rate; the existence of the thin fast etch region is noted as the front passes through.
- **Direct (No extension)**: Speeds are defined only in terms of the value of the etch function at each grid point. Thus, if grid points near the annulus are used, they may provide speed values which do not accurately reflect the local etch rate at the interface itself.

The results of the simulations for a 30×30 grid are shown in Figure 12.20. On the left, the direct solution method is shown; on the right, the extension technique is used. The underlying grid is indicated by the dotted lines. On the left side, considerable jaggedness occurs, since the effective speed of the front at each point will be an average of the speed at the surrounding grid points. A much smoother and more accurate evolving profile is obtained using the extension technique shown in the figures on the right. The extension velocity allows two fronts to come together and merge, as expected.

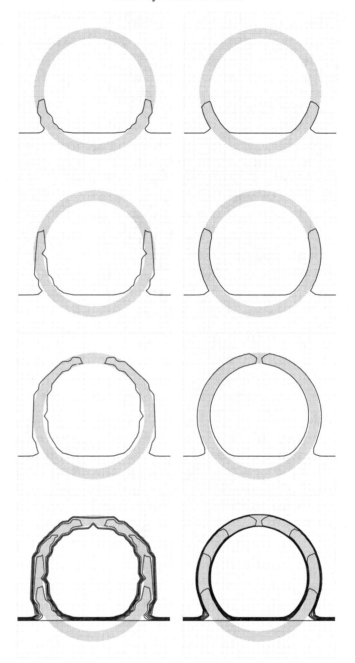

Fig. 12.20. Left = No extension. Right = Extension method.

13

Building Level Set and Fast Marching Applications

Outline: *To end this part, we provide some guidelines about how to implement Level Set and Fast Marching methods.*

When all is said and done, how does one actually **implement** Level Set Methods and Fast Marching Methods in complex physical models? First, we stress the obvious. In order to devise either a level set or Fast Marching simulation, one must have a good model for the interface velocity. The numerical algorithms presented in this book say how to move an interface with a prescribed speed; they assume that one already knows how to choose that velocity. A good numerical algorithm efficiently (and, if possible, elegantly) does what it is told to do. But it cannot (and should not) make modeling decisions about physics.

With that said, we return to the level set formulation, namely

$$\phi_t + F_{\text{ext}} |\nabla \phi| = 0.$$

As we shall see throughout the applications which follows, this speed F_{ext} can depend on a wide collection of terms, including such factors as the local curvature, the normal to the curve or surface, passive advection fields, and solutions to partial differential equations with jump conditions and/or source terms determined by the interface location. We can make a somewhat abstract categorization of the terms which can influence the speed function F.

161

- F^{Natural}: These are terms that depend on local shape properties of the interface, such as the curvature, second derivative of the curvature, normal direction, etc. As such, they have a natural definition, not just for the zero level set but for all the level sets. Thus, we can easily build the extension velocity $F_{\text{ext}}^{\text{Natural}}$ by simply evaluating this component at each grid point using the level set passing through that grid point. Other examples of such influences include passive advection velocities, that is, those which depend on the location in the domain.

- $F^{\text{Restricted}}$: These are terms that have meaning only on the front itself. An example from semiconductor simulations is the determination of visibility of each point on the interface with respect to some outside source. In this case, we suggest using the extension methodology of Chapter 11 to build a speed function $F_{\text{ext}}^{\text{Restricted}}$ at each grid point in the Narrow Band.

- F^{Physics}: These are terms that come from the specific location of the interface, and they typically require solving an associated partial differential equation with jump conditions, boundary conditions, or source terms which depend on the interface location. We can write

$$\phi_t + F_{\text{ext}}^{\text{Physics}}(\phi = 0)|\nabla\phi| = 0,$$

where the expression $F_{\text{ext}}^{\text{Physics}}(\phi = 0)$ is meant to signify that the velocity depends on the particular location of the zero level set. In such cases, the first step is use the position of the interface as input to the appropriate partial differential equation. The next step is to solve this equation. The final step is the construction of the extension velocity. Typically, there are two choices. If the solution of the partial differential equation gives a velocity field everywhere, one might try to use that as the extension velocity. On the other hand, this can cause level set lines to stretch and bunch. Our experience indicates that evaluating this velocity field on the zero level set to produce a velocity on the front, and then using the extension velocity algorithm discussed earlier will provide the smoothest approach. Using this extension velocity field, and re-initializing only sparingly, if at all, helps maintain mass conservation and numerical accuracy.

For Fast Marching Methods, the key is to find a causal update algorithm which satisfies the viscosity framework. For the Eikonal equation, this has been presented in detail earlier. The scheme must have the causality feature that one need not go back and change a known value, and the approximation must converge to the given static Hamilton-Jacobi equation. An appropriate and viable update strategy for a slightly more complex static Hamilton–Jacobi equation is given in the applications section. In general, this is the most challenging part of applying the Fast Marching Method.

In the next part of the book, we discuss a variety of applications which employ level set methods and Fast Marching Methods. Throughout, we shall note the construction of extension velocities, alternative ways to formulate problems, and further tricks of the trade.

Part IV

Applications

In this part, we present a series of applications of Fast Marching Methods and Level Set Methods. The intent is to indicate the range of problems that may be framed in this perspective.

14

Geometry:
Curve/Surface Shrinkage and
Self-similar Surfaces

Outline: *We begin with the application of Level Set Methods to problems in the geometric evolution of curves and surfaces. The motion here depends only on geometric properties such as normal direction and curvature; nonetheless, this is a rich area.*

14.1 Statement of problem

Given a hypersurface in R^n propagating with some speed $F(\kappa)$, we have previously considered speed functions of the form $F(\kappa) = 1 - \epsilon\kappa$, where κ is the curvature. We now focus on the speed function $F = -\kappa$.

Why does one care about such a speed function? One answer is that this motion corresponds to a geometric version of the heat equation. Substituting the speed function into the level set equation (Eqn. 1.5) we have

$$\phi_t = \left[\nabla \cdot \frac{\nabla\phi}{|\nabla\phi|}\right] |\nabla\phi|, \qquad (14.1)$$

where we have substituted $\nabla \cdot \frac{\nabla\phi}{|\nabla\phi|}$ for the curvature κ. Thus, motion under curvature resembles a non-linear heat equation; large oscillations are immediately smoothed out, and long-term solutions result from dissipation of information about the initial state.

Another reason, as we shall see in later sections, is that motion by curvature plays an important role in many applications. Because it corresponds to a diffusion-like term, it can be used to relax and reshape boundaries, find minimal energy configurations, account for surface tension in flexible membranes, and act as a viscous term in physical phenomena.

Thus, as a precursor to a collection of problems, we investigate motion

167

of an interface under its own curvature. We would like to study several questions:

(i) What happens to a simple closed curve moving under its curvature?

(ii) What happens to a hypersurface in higher dimensions?

(iii) Are there self-similar structures under this motion, that is, hypersurfaces that do not change shape as they evolve?

(iv) What happens to surfaces moving under geometric flows determined by more general metrics?

(v) What happens to modifications of curvature flow that preserve the enclosed area?

(vi) What happens to curves moving under higher derivatives of curvature?

(vii) What happens with multiple regions, that is, the motion of triple points under curvature flow, and flows of multiple surfaces?

The first question about the evolution of a simple closed curve moving under its curvature was definitively answered through the remarkable work of Gage and Grayson. First, Gage [97, 98] showed that any convex curve moving under such a motion remains convex and must shrink to a point. Grayson [105] followed this work with the beautiful result that *all* simple closed curves must shrink to a point, regardless of their initial shape.

What about the second question? We first note that there are several different curvatures for hypersurfaces in R^3 and higher, including the mean curvature and Gaussian curvature. Focusing on mean curvature, Huisken [117] has shown that convex shapes shrink to spheres as they collapse, analogous to the result of Gage. However, Grayson [106] showed that non-convex shapes may in fact *not* shrink to a sphere and provided the counterexample of the dumbbell. A narrow handle of a dumbbell may have such a small inner radius that the mean curvature of the saddle point at the neck may still be positive, and hence the neck will pinch off. A study of motion under Gaussian curvature may be found in [185].

Next, what about self-similar shapes? In two dimensions, it is clear that a circle collapsing under its own curvature remains a circle; this can be seen by integrating the ordinary differential equation for the changing radius. In three dimensions, a sphere is self-similar under mean curvature flow, since its curvature is always constant. Are there other shapes?

Looking at variants of curvature flow, how general is the level set framework? For flows such as those which preserve enclosed area, how well do these numerical methods perform?

Next, what about flow under speeds that depend on derivatives of the curvature? Such flows are important in building elastica, in shape segmentation in computer vision, in surface diffusion in metal reflow in semiconductor manufacturing, and in the sintering of materials. Are there similar results for these problems?

Finally, the motions of multiple surfaces require a new formulation of the level set perspective. As discussed earlier, the standard level set formulation of [187] assumes that the interface is a hypersurface with a clear notion of inside and outside. When three or more separate regions come together at a single point, how can one devise techniques to model curvature flow?

14.2 Equations of motion

The transformation of these questions to a level set framework is particularly straightforward. Each level set moves under its own curvature, and the level set equation (Eqn. 14.1) is solved everywhere. Thus, the problem we "care" about is embedded in an entire family of such problems, each of which proceeds according to the prescribed motion.

14.3 Results

We begin with a calculation designed to illustrate Grayson's theorem. In Figure 14.1, an odd-shaped initial curve is viewed as the zero level set of a function defined on all of R^2. Here, for illustration, black corresponds to $\phi < 0$, while white corresponds to $\phi > 0$, and the zero level set is the boundary between the two. As the level curves flow under curvature, the ensuing motion carries each to a circle that then disappears. In the evolution of the front, one clearly sees that the large oscillations disappear quickly. As the front becomes circular, the motion slows, and the front eventually disappears.

Next, Grayson's counterexample of the collapse of a dumbbell in three dimensions under its mean curvature is computed. In Figure 14.2, taken from Sethian [228], the two-dimensional cross section of the motion of a dumbbell collapsing under its mean curvature is shown at various times. Because the principal radius of curvature around the neck is so small,

Fig. 14.1. $F(\kappa) = -\kappa$.

it dominates the negative value of the other principal radius, and the handle shrinks inward and disappears.

This result shows that a simply connected surface propagating under its mean curvature in three dimensions can break into two separate surfaces. In fact, a different choice of initial surface shows that the front can go from a single surface to *five* separate surfaces before each collapses to a sphere and disappears. Figure 14.3, taken from Chopp and Sethian [61], shows two connected dumbbells. As the handles collapse, the necks break off and leave a remaining pillow-like region behind. This pillow region collapses as well, and eventually all five regions disappear.

Finally, what about the existence of self-similar surfaces that remain morphologically unchanged as they evolve under curvature? Obviously, a sphere is one such shape. To find another, imagine a torus-shaped

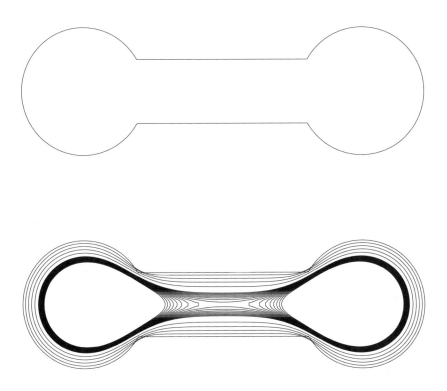

Fig. 14.2. Cross section of three-dimensional dumbbell evolution. Top: initial shape. Bottom: evolution.

object. If the torus is fat, then the inside will pinch off and become sphere-like; the shape will then continue to evolve and collapse to a point. On the other hand, if the initial shape is closer to a thin ring, the ring will collapse down and disappear all at once. This suggests the existence of an intermediate shape between the two that exactly balances these two competing pulls so that the shape stays the same as the object collapses. In fact, Angenent [13] was the first to prove the existence of a self-similar torus that preserves this balance.

Are there other self-similar shapes? Here, we describe an algorithm due to Chopp that yields a collection of such surfaces. The following

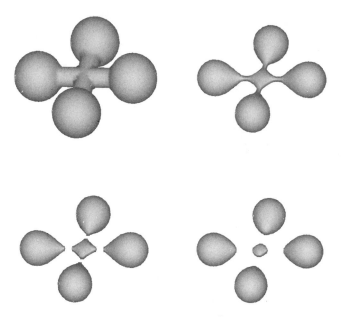

Fig. 14.3. Collapse of two-handled dumbbell.

discussion is taken from Chopp [60]; we refer the interested reader to that work.

In order to produce self-similar shapes, we need two things. First, since hypersurfaces get smaller as they move under their curvature, a mechanism is needed to "rescale" their motion so that the evolution can be continued towards a possible self-similar shape. Second, a way of testing self-similarity is required.

Chopp first constructed an evolution equation associated with rescaling. Begin with the level set equation for mean curvature flow, given by

$$\phi_t = \kappa |\nabla \phi|. \tag{14.2}$$

Then define a new function ψ as

$$\psi(x,t) = \phi(\sigma(t)x, t)/\sigma^2(t) - L(t), \tag{14.3}$$

where $\sigma(t)$ acts as a stretching function, and $L(t)$ will be used to pick out the level set corresponding to the front of interest. Differentiating Eqn. 14.3 with respect to t and combining with Eqn. 14.2 produces a new partial differential equation for ψ given by

$$\psi_t = \frac{\sigma'(t)}{\sigma(t)} \left(x \cdot \nabla \psi - 2(\psi + L(t)) \right) + \kappa_\psi ||\nabla \psi|| - L'(t), \qquad (14.4)$$

where κ_ψ is the mean curvature of the level set surface of ψ through the point (x, t).

The functions $\sigma(t)$ and $L(t)$ are determined dynamically as time evolves. The stretching function $\sigma(t)$ is chosen so that the zero level set of ψ has constant volume for all t. Assuming that the interior of a surface is given by $\phi(x, t) < 0$, we define

$$V(t) = \int_{\{x : \phi(x,t) \leq 0\}} dV.$$

Then

$$\sigma(t) = \left(\frac{V(t)}{V(0)} \right)^{1/3},$$

so that

$$\frac{\sigma'(t)}{\sigma(t)} = \frac{V'(t)}{3V(t)}.$$

Note that the function $V'(t)/3V(t)$ is independent of scale. This scale-independent function can be evaluated using ψ instead of ϕ. It is approximated numerically by assuming $\tilde{\phi}(x, t) = \psi(x, t)$, and then letting $\tilde{\phi}(x, t)$ flow by mean curvature alone (i.e., σ remains fixed) to $\tilde{\phi}(x, t+\Delta t)$. The change of volume can be used to compute $V'(t)/3V(t)$.

The level function $L(t)$ is determined by using a predetermined function $\rho(\ell, t)$ that depends solely on the geometry of the level surface $\phi(x, t) = \ell$. In the case of the torus, ρ measures a ratio of the inner radius of the torus to the thickness of the cross section of the torus. In this way, $\rho \to 0$ if the torus evolves toward a sphere, and $\rho \to +\infty$ if the torus evolves toward a thin ring. Furthermore, a self-similar torus would necessarily have $\frac{\partial \rho}{\partial t} = 0$, because the geometric ratio should remain constant. The level function $L(t)$ is then defined by the equation $\frac{\partial \rho}{\partial t}(L(t), t) = 0$. In this context, the self-similar solution is the limiting level surface $L(t)$ of $\psi(x, t)$. For details, see [60].

How does one check that the surfaces so obtained are indeed self-similar? Here, one can use a simple condition given by Huisken [118] that is sufficient for self-similarity. Suppose that Γ_0 is a 2-manifold such that for each point $x \in \Gamma_0$,

$$\kappa(x) = (2T)^{-1/2} x \cdot n, \qquad (14.5)$$

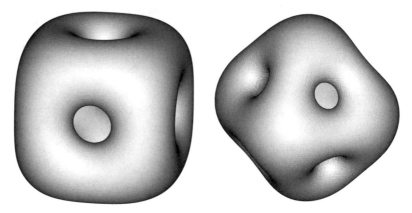

(a) Self-similar cube with holes   (b) Self-similar octahedron with holes

Fig. 14.4. Self-similar shapes.

where n is the unit normal to Γ_0 at x and T is some constant. Define $\Gamma_t = (2(T - t))^{1/2}\Gamma_0$. Then

$$\left(\frac{d}{dt}\Gamma_t\right)^{\perp} = \frac{-1}{(2(T-t))^{1/2}}\kappa_0 \cdot n = -\kappa n. \qquad (14.6)$$

Therefore, Γ_t is equivalent to curvature flow with initial surface Γ_0, and hence Γ_0 must be a self-similar solution. The constant T is the time to singularity for the surface Γ_0. This test can be implemented numerically in a fairly straightforward manner.

Using the above technique, Chopp has identified a series of self-similar surfaces. One family of surfaces is constructed as follows. Begin with a sphere, and drill a hole in each of the three coordinate directions. One can think of each hole as exiting the face of a cube, producing a surface of genus five. As this surface evolves under mean curvature flow, the holes expand and contort. The final shape, shown in Figure 14.4(a), is numerically self-similar.

This construction continues through Chopp's observation that "every regular polyhedron with holes will produce a corresponding self-similar solution for mean curvature flow" (see [60]). Figure 14.4(b) shows another surface, produced by starting with a sphere with six cylinders drilled into it through the diagonals. This is equivalent to an octahedron, and yields a surface of genus seven. This construction of regular polyhedra of higher and higher order yields a family of self-similar surfaces.

The above technique produces one class of self-similar surfaces. Chopp has also constructed a second class of unbounded self-similar surfaces using a slightly different ρ function. This second class revealed further theoretical results; for example, see Angenent, Ilmanen and Chopp [14]. These and other such pictures may be found in Chopp [60].

14.4 Flows under more general metrics

As discussed earlier, the fundamental level set embedding approach can be extended to complex flows under more general metrics. Pasch [190] has concentrated on a Schwarzschild-type metric and produced some very nice results. First, Pasch deals with the inevitable small time step problem inherent in parabolic problems by using an exponential integrator. He considers and compares two options. First, he uses an integrator which uses numerical differentiation to extract directional derivatives. Next, he uses one which uses the exact Jacobian for the derivative evaluation. The two competing techniques are carefully compared.

Pasch looks at several different flows. He begins with a study of the classical Schwarzschild model, using a metric on the time slice $\tau = 0$ of the form

$$g_{ij}(p) = \delta_{ij} \left(1 + \frac{m}{2r} \right)^4 ,$$

with $r = |p|$. This metric corresponds to a stable minimal surface around the singularity, representing the horizon with a static hole in general relativity. For $m = 1$, the horizon is the sphere around zero with radius $1/2$. Pasch studies flows starting with an initial torus, and the ensuing pinching and motion. He also studies neck formation starting from an initial ellipsoid and motion with under a three singularity flow. We refer the reader to [190] for these and further results.

14.5 Volume-preserving flows

We now consider a variant on the level set flow, given in [61], and examine $F = -(\kappa - \kappa_{\mathrm{Ave}})$, where κ is the curvature of the interface, and κ_{Ave} is the average curvature of the interface, defined as

$$\kappa_{\mathrm{Ave}} = \frac{\int_{\phi=0} \kappa ds}{\int_{\phi=0} ds} .$$

Initial area	Final area	Percentage change
0.50765	0.50410	.699%

Initial length	Final length	Percentage change
11.6723	2.51732	78.42 %

Fig. 14.5. Changes in area and length under constant volume flow.

Here, our notation means that the integrals are performed along the level set $\phi = 0$; s is arclength. The denominator is just the length of the curve; while the numerator is the integral of the curvature around the curve. We note that this is a variant on the level set flow for constructing surfaces of prescribed curvature given in [61]. The area enclosed by the interface under this flow remains constant, since we have that

$$\frac{d}{dt}\mathrm{Vol}(t) = \int_{\phi=0} F ds = - \int_{\phi=0}(\kappa - \kappa_{\mathrm{Ave}}) ds = -\left(\int_{\phi=0} \kappa ds - \kappa_{\mathrm{Ave}} \int_{\phi=0} ds \right) = 0. \tag{14.7}$$

Numerically, we must calculate the length and the integral of the curvature. In each cell containing the zero level set we find the interface and then perform integration in that cell. Finding such cells requires only a quick sweep through the narrow band. Whenever quantities such as curvature are needed, we interpolate those values from the mesh. Summing up over such cell contributions provides the desired integrated quantities.[1]

In Figure 14.5 and Figure 14.6 we show this constant volume flow applied to the "blob" studied earlier for curvature flow. In Figure 14.5, we show the change in area vs. the change in length. As the calculation proceeds, the results show that area is conserved even for such a wildly distorted initial shape. In 14.6 we show the evolving flow; the calculation is not run to its final circular state.

---

[1] We note that in this, as well as some other applications, we actually construct the zero level set in each cell as part of building the front velocity. There is nothing intrinsically wrong or undesirable with this approach. Our partial differential equations view of evolving interfaces often uses this construction. The key is to avoid *moving* this constructed representation.

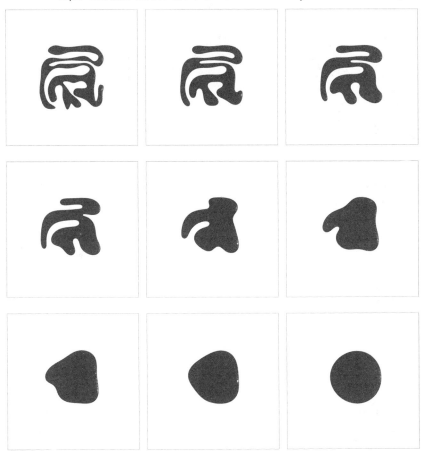

Fig. 14.6. $F(\kappa) = -(\kappa - \kappa_{\mathrm{Ave}})$.

14.6 Motion under the second derivative of curvature:
surface diffusion

The methodology can be extended to another practical problem, which, while still purely geometrical, pushes our approach quite a bit further. Sintering (see [141, 195, 275]) is the process under which a compact consisting of many particles is heated to such a high temperature that the particles become a viscous creeping fluid, and the particles begin to coalesce together. One of the oldest technological examples involves the production of bricks; other examples include formation of rock strata from sandy sediments and the motion of thin films of metals in the microfabrication of electronic components.

Iter=1   Iter=100   Iter=200

Fig. 14.7. Motion of ellipse under speed $F = \kappa_{\alpha\alpha}$.

At issue is the solution of the equations for creeping flow, in which the body forces on the boundary of the materials depend on the tangential stress derivative on the boundary. In one component of this model, the interface speed F in its normal direction depends on the second derivative of the curvature, where the derivatives are taken with respect to arc length α. Thus, in our level set framework, one wants (in two dimensions) to follow a curve propagating with speed $F = -\epsilon\kappa_{\alpha\alpha}$. Thus,

$$\phi_t + \epsilon\kappa_{\alpha\alpha}|\nabla\phi| = \phi_t + \epsilon\left[\nabla \cdot \frac{\nabla\phi}{|\nabla\phi|}\right]_{\alpha\alpha} = 0,$$

$$\phi_t = -\epsilon\nabla\left[\nabla\left[\nabla \cdot \frac{\nabla\phi}{|\nabla\phi|}\right] \cdot \frac{(\phi_y, -\phi_x)}{|\nabla\phi|}\right] \cdot \frac{(\phi_y, -\phi_x)}{|\nabla\phi|}. \tag{14.8}$$

This means that the speed depends on the *fourth derivative* of the level set function. We immediately note that a circle is a stable object, since the curvature is constant. A little examination leads one to think that an ellipse undergoes a restoring force which brings it back into a circle.

What about more complex shapes? The problem is quite subtle. Numerical experiments are notoriously unstable when they involve computing fourth derivatives[2], and are eloquently described by Van de Vorst [275]; he uses marker particle schemes together with elaborate remeshing strategies to keep the calculation alive.

A level set approach to this problem was developed by Chopp and Sethian [62]. In that work, the individual derivatives in the above expression were approximated by central difference approximations and are used to study the motion of a sequence of closed curves to analyze flow under the second derivative of curvature given by Eqn. 14.8. Here, we summarize some of the results in [62]. First, in Figure 14.7, we show the evolution of a simple ellipse under this motion. The transformation

[2] For example, computing the solution to the biharmonic equation is delicate.

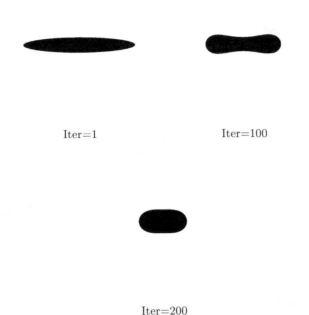

Fig. 14.8. Motion of elongated ellipse under speed $F = \kappa_{\alpha\alpha}$.

shows the elliptical initial state on the left, followed by the evolution into a circle, which then remains fixed after a large number of calculations.

This might seem to indicate that a convex shape remains convex as it flows under this evolution equation. This in fact is not true, as seen by examining the motion of a slightly more elongated ellipse in Figure 14.8. At points of high curvature, the interface moves inward, leaving a bulge which propagates around to the flatter sides until the interface balances itself out. In the case of sharp corners, the effect is more pronounced, as seen in Figure 14.9. Figure 14.9 shows the evolution of several non-convex initial shapes, all of which approach the stable state of a circle. The curves are shown at uneven times, and the flows are not completed.

Second derivative flow becomes even more murky in the face of topological change. Imagine two ellipses, each with a large ratio between the major and minor axes. If they are put side by side (rather than end to end), the flatter sides will cross over each other, and one expects

Fig. 14.9. Motion of non-convex curves under speed $F = \kappa_{\alpha\alpha}$.

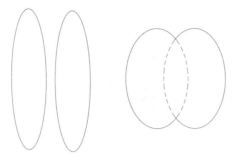

Fig. 14.10. Motions of two ellipses under the second derivative of curvature.

(at least in many physical situations) some sort of merger, as in Figure 14.10.

This example underlies the difference between curvature flow ($F = -\kappa$) and flow by the second derivative of curvature. In the former, a maximum principle ensures that two separate closed curves will always remain separate under this flow. This allows the sort of natural embedding prescribed by a level set interpretation. In contrast, flow by the second derivative of curvature has no such property, as demonstrated in the previous example. Hence, the notion of embedding the motion of the two ellipses in a single "level set function" requires thought.

In order to execute true merger of two regions moving under the second derivative of curvature, care must be taken. In Fig. 14.11, taken from [62], we show the merger of two rectangular regions. The results show how the regions come together. Finally, in Figure 14.12, also taken from [62], we show an example of three-dimensional flow under the Laplacian of curvature, revealing the smoothing effects of this flow. For details, see [62].

Finally, we note a recent paper on the topic of motion under second derivative of curvature flow by Escher, Mayer, and Simonett [84]. On the theoretical side, in any number of dimensions, they prove that the area of the bounding interface is non-decreasing under this flow and that the enclosed volume remains constant. They also discuss the existence and uniqueness of solutions to this flow. In addition, they provide a series of numerical experiments using a marker particle approach showing the motion of a figure-eight, as well as the motion of a single closed curve that crosses over itself during its evolution and then returns to being a simple curve.

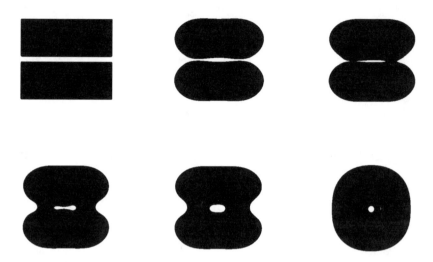

Fig. 14.11. Merger of two separate regions under $F = \kappa_{\alpha\alpha}$.

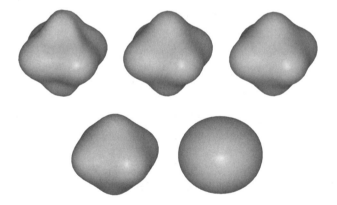

Fig. 14.12. Flow under Laplacian of Curvature

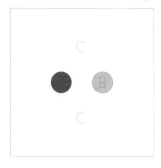

Fig. 14.13. Regions A and B expand into region C.

14.7 Triple points: variational and diffusion methods

14.7.1 Statement of problem

As initially designed in [187], the level set technique applies to problems in which there is a clear distinction between an "inside" and "outside." This is because the interface is the transition between positive and negative values for the level set function. Extensions to multiple (more than two) interfaces have been made in some specific cases. In the case in which interfaces are passively transported and behave nicely, one may be able to use only one level set function and judiciously assign different values at the interfaces. For example, the zero level set may correspond to the boundary between two regions A and B, with the level set value 10 corresponding to the interface between two regions B and C. If A and C never touch, then this technique may be used to follow the interfaces in some cases.

However, in the general case involving the emergence and motion of triple points, a different approach is required, since many different situations can occur (see, for example, Bronsard and Wetton [40] and Taylor, Cahn, and Handwerker [263]). Consider the following example, as illustrated in Figure 14.13. Regions A and B are both circular disks growing into region C with speed unity in the direction normal to each interface. At some point, the interfaces will touch and meet at a triple point, where a clear notion of "inside" and "outside" cannot be assigned in a consistent manner.

One solution lies in recasting the interface motion as the motion of one level set function for each material. We now discuss a technique, presented in [231, 232], which works in some cases, but is by no means general enough to handle full complexity.

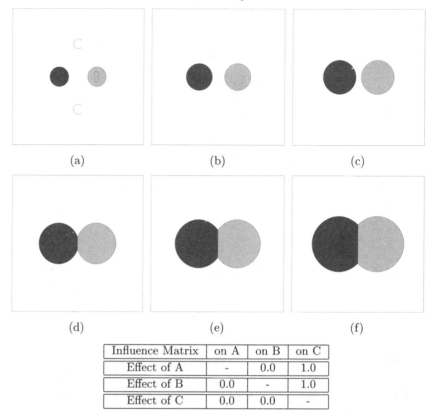

Influence Matrix	on A	on B	on C
Effect of A	-	0.0	1.0
Effect of B	0.0	-	1.0
Effect of C	0.0	0.0	-

Fig. 14.14. A and B move into C with speed 1, stop at each other

Imagine N separate regions and a full set of all possible pairwise speed functions F_{IJ} that describe the propagation speed of region I into region J: F is taken as zero if region I cannot penetrate J. The idea is to advance each interface to obtain a trial value for each interface with respect to motion into every other region, and then combine the trial values to obtain the maximum possible motion of the interface.

In general, then, we proceed as follows. Given a region I, obtain $N-1$ trial level set functions ϕ_{IJ}^* by moving the region I into each possible region J, $J = 1, \ldots, N$ $(J \neq I)$ with speed F_{IJ}. During the motion of region I into region J, assume that all other regions are impenetrable, that is, use the masking rule given by Eqn. (12.1). We then test the penetrability of region J itself, leaving the value of ϕ_{IJ}^* unchanged if $F_{IJ} \neq 0$, else modifying it with the maximum of itself and $-\phi_{IJ}^*$. Finally,

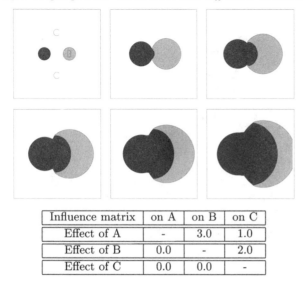

Influence matrix	on A	on B	on C
Effect of A	-	3.0	1.0
Effect of B	0.0	-	2.0
Effect of C	0.0	0.0	-

Fig. 14.15. A into C with speed 1, A in B with speed 3, B in C with speed 2.

to allow region I to evolve as much as possible, we take the minimum over all possible motions to obtain the new position; this is the re-ignition idea described earlier. Complete details may be found in Sethian [231].

Three examples are shown to illustrate this approach. Given regions A, B, and C, the *influence matrix* describes the interactions between each pair of regions. The interaction of each region with itself is left blank. The interaction of any pair of regions must be zero in one of the two interactions.

In Figure 14.14, regions A and B expand with unit speed into region C, but cannot penetrate each other. They advance and meet; the boundary between the two becomes a vertical straight line.

Next, we consider a problem with different evolution rates. In Figure 14.15, region A grows with speed 1 into region C (and region C grows with speed 0 into region A), and region B grows with speed 2 into region C. Once they come into contact, region A dominates region B with speed 3, thus region B grows through C and then is "eaten up" by the advancing region A. Note what happens. Region A advances with speed 3 to the edge of region B, which is advancing only with speed 2 into region C.

Finally, in Figure 14.16 the motion of a triple point between regions A, B, and C is shown. Assume that region A penetrates B with speed 1, B

Geometry

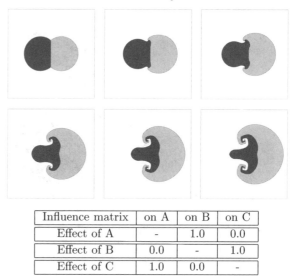

Influence matrix	on A	on B	on C
Effect of A	-	1.0	0.0
Effect of B	0.0	-	1.0
Effect of C	1.0	0.0	-

Fig. 14.16. Spiraling triple point: 98 × 98 grid.

penetrates C with speed 1, and C penetrates A with speed 1. The exact solution is given by a spiral with no limiting tangent angle as the triple point is approached. The triple point itself does not move. In Figure 14.16, results are shown from a calculation on a 98 × 98 grid. Starting from the initial configuration, the regions spiral around each other, with the leading tip of each spiral controlled by the grid size. In other words, we are unable to resolve spirals tighter than the grid size, which therefore controls the fine scale description of the motion. However, we note that the triple point remains fixed. A series of additional calculations using this approach may be found in [232].

Several words of caution are in order. This approach is expensive, in that one considers the interaction of each region with every other region. In some particular cases, significantly fewer interactions are required (see for example, [5]), which makes this a valuable approach. Regardless of the computational speed involved, an additional complexity may stem from difficulty in constructing the details of the influence matrix.

14.7.2 More complex motions

The true motion of multiple interfaces and multiple junctions is, in many cases, highly complex. As illustration, imagine a triple point, in which

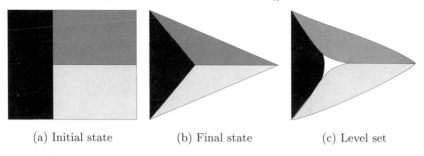

(a) Initial state   (b) Final state   (c) Level set

Fig. 14.17. Evolution of triple point under curvature.

each of the three regions is attempting to move according to its own curvature. In Figure 14.17(a), we show an initial configuration, and in Figure 14.17(b) we show the final configuration, which consists of the three lines meeting in equal angles of 120 degrees.

If one attempts to apply the level set method for multiple interfaces described in the previous section, a difficulty occurs because each level function attempts to move away from the others, creating a gap.

To see this more clearly, we consider the motion of three interfaces under curvature as in Figure 14.18a. We draw end points for each of the curves to give boundary conditions, although we shall think of these curves as stretching off to infinity. The first question to deal with is the meaning of "curvature" at the triple point. This issue has been discussed in detail in many papers (see, for example, [199, 263]), and we refer the reader there for some theoretical discussions. Following these views, we evaluate the curvature from each side as determined by the region's boundary, and then view curvature flow as the net effect of all these forces.

It seems clear from an energy argument that the final configuration should be a triple point in which the angles are all 120 degrees, as shown in Figure 14.18b. However, imagine one were to apply a straightforward level set method to this problem, using one level set function for each region, negative inside and positive outside. As discussed previously, each level set function would evaluate the curvature for its own level set representation, and the triple point would "pull apart". In Figure 14.17(c), we show this gap developing when a level set technique is applied to the final state.

To tackle these and similar problems, two techniques for curvature-driven flow were proposed in a paper by Merriman, Bence and Osher [176]. This was followed by the "variational level set method" given in

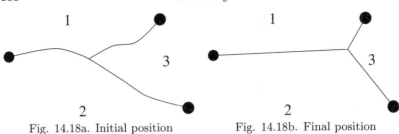

Fig. 14.18a. Initial position   Fig. 14.18b. Final position

Fig. 14.18. Motion of triple point under curvature.

[282], which provides a nice extension of level set methods to problems of multiple interfaces, based on the theoretical work of [199], and also by Ruuth's fast spectral-based approach [211]. We now discuss some of these ideas, and refer the interested reader to the cited papers for more details and a large number of sample calculations and convergence tests.

14.7.3 Diffusion-generation curvature motion

The first scheme given by Merriman, Bence, and Osher in [176] is as follows. Suppose one characterizes each region i by a characteristic function χ_i and labels this initial position $\chi_i(t = 0)$. Then imagine diffusing each region by the heat equation for time step Δt to obtain temporary values $\chi_i^*(t = \Delta t)$. This can be thought of as convolution with a Gaussian distribution. One can then rebuild sharp characteristic sets by letting the biggest value of these temporary values determine the region that owns a point. In other words, the new characteristic function for each region i is given by

$$\chi_i(t = \Delta t) = \left\{ x, y \,|\, \chi_i^*(t = \Delta t)(x,y) > \chi_j^*(t = \Delta t)(x,y) \right\}$$

where the j index represents testing over the other characteristic functions. One can thus think of this algorithm as diffusing a little, sharpening things up, and then repeating.

   This is an appealing algorithm. To begin, although the algorithm is linear, it approximates curvature flow, which is a non-linear equation. It handles topological merger and works in both two and three dimensions. A series of impressive calculations of triple points moving under curvature flow was presented in [176]; theoretical proofs about the convergence of this technique to curvature flow were presented in [21, 85].

   The algorithm used to approximate the diffusion step in the above was a finite-difference based solution to the heat equation. An alternative approach, based on a spectral approximation to the Gaussian smoothing and resulting in a fast technique, was developed by Ruuth in [211]; some recent work in [212] carries these ideas even further. We refer the interested reader to these papers, as well as to continued work on extending the diffusion approach to more complex curvature-driven motion and other works on curvature flow using level set methods (see, for example, [179]).

14.7.4 Constraint-based level set flows

A different algorithm, also posed in [176], tries to explicitly deal with the fact that the level set functions pull apart when viewing each region as its own level set problem. In this technique, one directly confronts the developing gap and resets the level set functions to close this gap. This is done through enforcing a constraint condition at the end of each time step as follows. First, given the level set functions ϕ_i^n at time n, where the index i corresponds to the different level set functions, compute temporary new level set functions called ϕ_i^* at time $(n+1)\Delta t$, and then reset them by

$$\phi_i^{n+1} = \phi_i^* - \max_{i \neq j} \phi_j^*.$$

This keeps the triple point in place; however, the cost is that the level set functions can develop spontaneous zero crossings later in time. A remedy is to re-initialize all the level sets using any of the re-initialization techniques described in the previous chapter.

A more sophisticated version of this technique, which relies on the theoretical formalisms developed by Reitich and Soner [199] and allows more complex energy balances between the competing regions, was developed by Zhao, Chan, Merriman and Osher in [282]. The idea is to define an energy of the combined set of interfaces, and then to minimize this energy subject to the constraint equation

$$\sum_{i=1}^{n} H(\phi_i(x,y)) - 1 = 0.$$

This constraint equation keeps the level sets together without overlapping. In fact, rather than solve this set of constraints, they replace it by a single least squares constraint equation. The energy minimizer is re-interpreted as a level set flow problem, in which the amount of overlap and/or vacuum between level sets helps drive the sets either together or apart as part of the energy-minimizing motion. This technique limits angles to the classical condition of energy balance at triple points dependent on surface tension ratios, and is not straightforward in the case of many distinct regions. Nonetheless, it has proved quite powerful as a way of computing a wide collection of steady-state energy minimizing bubble configurations, and we refer the interested reader to [282, 283].

15

Grid Generation

Outline: *We continue with the "pure" geometry problem of grid generation. The goal is to create a logically rectangular, body-fitted grid, that is, a grid such that each grid point has four (six) neighbors in two dimensions (three dimensions), and a set of grid lines corresponds to the body. If we view the body itself as an interface and advancing the interface away from the body, a level set framework provides a natural grid construction.*

15.1 Statement of problem

Imagine that one is given a closed body, either as a curve in two space dimensions or as a surface in three space dimensions. In many situations, one wishes to generate a logically rectangular, body-fitted grid around or inside this body. By "logically rectangular" we mean that each node of the grid has four neighbors (in two dimensions; in three dimensions there are six neighbors); by "body-fitted", we mean that the grid aligns itself with the body so that one set of coordinate lines matches the body itself. Such grids are of importance in a variety of areas. For example, in computational fluid dynamics, one might desire a finite difference approximation at each grid point to the Navier–Stokes equations that describe fluid flow around the body. Often, the solution requires that the fluid velocity vanish in the direction both normal and tangential to the solid body. A coordinate system aligned with the body can make the imposition of these boundary conditions considerably easier.

This problem of *grid generation* is difficult, in part because of two competing desires: (1) cell uniformity, in which all grid cells are of approximately the same area, and (2) mesh orthogonality, which contributes to the accuracy of the finite difference approximation. Only in very special

191

cases can both be obtained. The science and art lie in balancing these two goals.

The subject of grid generation is vast, and there are many competing techniques. The reader is referred to the books by Knupp and Steinberg [140] and Castillo [52], as well as the work of Eiseman [81] and Thompson et al. [267]. Of course, rectangular grids are not the only possibility; unstructured grids, in which points and their neighbors are less regularly linked, play important roles in a variety of applications. Nonetheless, the focus here will be on logically rectangular, structured grids that balance the competing demands of uniformity in cell sizes and mesh orthogonality.

Broadly speaking, techniques for constructing logically rectangular grids include the following:

- *Algebraic/approximation methods*: In these techniques, a grid of points is placed around the body such that one coordinate line of these points is fitted to the body. The points are then adjusted and fit according to a variety of approximation functions, including polynomial techniques and fitting functions.

- *Elliptic methods*: In these techniques, an elliptic partial differential equation is solved throughout the entire domain, often through iteration or direct techniques. Grid lines are particular solution values of the equation.

- *Hyperbolic techniques*: Here, a front is advanced away from the body in such a way as to produce an appropriate set of grid lines as the front propagates outward.

- *Variational methods*: The positions of the grid points are produced through minimizing certain functionals.

Each of the above techniques has virtues and drawbacks. In general, the difficult issues may be summarized as follows:

- Very close to the body, it can be difficult to construct cells that conform to the body landscape and describe a smooth transition into the flow region.

- In the neighborhood of sharp inward corners, grid lines have a tendency to cross over themselves or create cells with inappropriate aspect ratios.

- In the neighborhood of sharp outward cusps, such as in edges of airfoils, grid lines often yield highly uneven cells, especially in regions where high accuracy is desired.

- Gridding together multiple bodies is, for the most part, difficult and still an art. None of the above techniques can easily anticipate other bodies when creating grids, and stitching together individual grids for each region into a coherent whole can be problematic.

Level set techniques offer a method for tackling some of these issues. The idea, as presented in Sethian [230], is to exploit the geometric nature of the problem and view the body itself as the initial position of an interface that must be advanced outwards away from the body. The initial position of the interface and its positions at later times form one set of grid lines; its orthogonal set forms the other. This technique roughly falls into the category of a hyperbolic solver. However, by solving the correct evolution equation for an advancing front, the difficulties of shock formation and colliding characteristics that plague most hyperbolic techniques are avoided. User intervention is kept to a minimum. For the most part, grids are generated automatically without the need to adjust parameters.

15.2 Equations of motion

The basic philosophy is to view the body itself as the zero level set of a function, and advance the front away from the body, for either an interior or exterior grid, using a chosen speed function $F(\kappa)$. At discrete chosen time intervals, zero contours of the level set function ϕ are constructed and serve as one set of grid lines. Lines transverse to these grid lines are then constructed. The discussion begins with two-dimensional grids; three-dimensional grids follow later.

15.2.1 Construction of body-fitted lines

First, one must construct an appropriate speed function to generate body-fitted coordinate lines outside a given body. Previous analysis (see [225]) shows that a front propagating with speed $F(\kappa) = 1 - \epsilon\kappa$ evolves towards a circle and hence yields a far-field circular grid. The rate at which the evolving front becomes circular depends on the choice of ϵ; the larger the value of ϵ, the faster the decay in the variation of curvature around the front. In order to ensure that no points of initially high positive curvature move backward, a minimum positive threshold speed value is chosen. The choice of this threshold value depends on the initial node placement. If a rectangular far-field grid is desired, the developing circular grid can be blended with a distant rectangle; periodic grids are obtained from periodic boundary conditions.

With a different choice of speed function, interior grids may be generated. In the case of a convex initial shape, the front must always remain convex, and hence the speed function $F(\kappa) = -\kappa$ is effective. Once again, in the case of a non-convex initial body, points with high curvature (in this case, high negative curvature) can move against the flow of the grid; again, a remedy is to include a threshold value which ensures that the front always moves inward, i.e.,

$$F(\kappa) = \min(-\kappa, F_{\text{threshold}}). \tag{15.1}$$

For three-dimensional grids, we have seen that fronts may change topology as they evolve. However, in many examples where an interior grid is desired, the above speed function will yield flow toward a single point.

15.2.2 *Construction of transverse lines*

Given the above set of body-fitted field lines, transverse lines must be constructed. The goal is to balance the competing desire of orthogonality and equal area/size in the construction of the field lines. This construction of transverse lines produces a grid which may be later adjusted through a variety of elliptic and parabolic smoothers available from more traditional techniques.

The basic technique constructs field lines by following trajectories normal to outward-evolving grid lines. This corresponds to following the gradient of the level function ϕ. Place N nodes at points X_i, $i = 1, \ldots, N$, on the initial body, and solve the N ordinary differential equations

$$\frac{dX_i}{dt} = F(\kappa)\frac{\nabla\phi}{|\nabla\phi|}, \tag{15.2}$$

where $X_i(t)$ is the location of the ith grid point at time t; a second order (Heun's) method can be used.

For any choice of ϵ greater than zero, the transverse lines cannot intersect. However, for non-convex initial curves, the constructed transverse lines can come quite close together. The larger the choice of the smoothing parameter ϵ, the more the transverse gradient lines are kept apart, however, at a cost of a large separation of the body-fitted field lines. Figure 15.1 shows exterior grids generated above a periodic cosine front for various choices of the smoothing parameter. For ϵ small (Figure

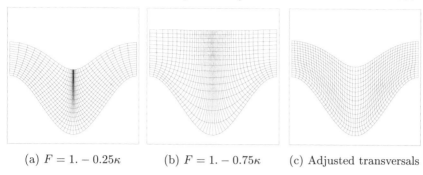

(a) $F = 1. - 0.25\kappa$ (b) $F = 1. - 0.75\kappa$ (c) Adjusted transversals

Fig. 15.1. Construction of transversals for propagating sine curve.

15.1(a)), the transverse lines come close together. For ϵ large (Figure 15.1(b)), transverse lines are kept apart, but the rapid motion of the evolving front leaves cells with large aspect ratios. Judicious choice of the smoothing parameter based on the initial node placement can solve some of these problems.

However, in the case of significant curvature variation in the initial body, some transversal adjustment is desirable. This is accomplished in two ways. First, transverse line trajectories may be readjusted so that their separation distance is proportional to the separation distance along the previous field line. Full use of this technique results in uniform cells; however, grid line orthogonality is not maintained. Second, transverse lines may be readjusted according to the length of the given field line; the total length of the field line is computed, and transverse nodes are readjusted to maintain equal spacings. Points where the derivative of the curvature vanishes, which may be calculated from a suitable difference expression acting on ϕ, act as stable points in this process and are not adjusted. Once again, this technique is linearly-weighted against the basic trajectory advection. In addition, a one-dimensional diffusion operator can be applied to the body-fitted field lines to separate transverse lines. Figure 15.1(c) shows the use of these techniques applied to the construction of a full set of grid lines above the cosine curve.

Finally, we note several details of the level set grid generation technique:

• Node placement on the boundary and the ensuing interior/exterior grid can be automatically controlled; placement can depend on local curvature, external functional dependence, or user-controlled specifications.

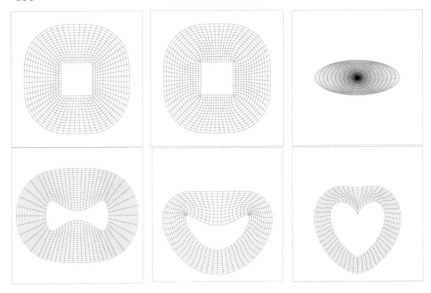

Fig. 15.2. Body-fitted grids generated using level set approach: I.

- In the case of corners and cusps in the initial boundaries, node placement may be altered to produce a rarefaction fan of smooth grid lines away from the singularities.
- In some cases, particularly with interior grids, some sort of refinement or coarsening is required; this can be accomplished by hierarchically subdividing or coarsening the grid.

For details, see [230].

15.3 Results, complications, and future work

A variety of grids can be generated using the approach outlined above; the following figures are all taken from Sethian [230]. In Figure 15.2, grids around some relatively straightforward objects are generated to show the idea behind the approach. In Figure 15.3, exterior and interior grids around significantly oscillatory objects are developed (note the ability of the method to resolve deep intrusions and extrusions). In Figure 15.4, grids around sharp objects are constructed, and in Figure 15.5, additional complex grids are produced.

Fig. 15.3. Body-fitted grids generated using level set approach: II.

Several improvements to the above techniques are desired. First, in the case of highly non-convex bodies, the above techniques will not work: sides of the body will grow together before the front has "time" to escape. In this case, some sort of domain decomposition is required. Second, the use of more standard techniques for modifying the grid, once the basic design is achieved, might prove fruitful. In particular, the grids generated using this approach may make excellent initial grids for variational and algebraic methods. Finally, extension of this work to multiple bodies requires hybrid techniques, similar to those commonly used in other

Grid Generation

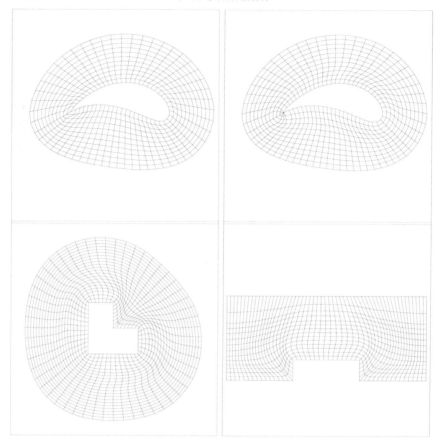

Fig. 15.4. Body-fitted grids generated using level set approach: III.

schemes. in which single grids are patched together or allowed to over-lap. An attractive approach currently under study is the use of these level set techniques for grids near bodies, coupled with a transition to Cartesian grids away from the body; this may be of particular use in fluid dynamics calculations where body-fitted coordinates are attractive to resolve boundary layer calculations, while far-field Cartesian grids are appropriate to connect complex geometries.

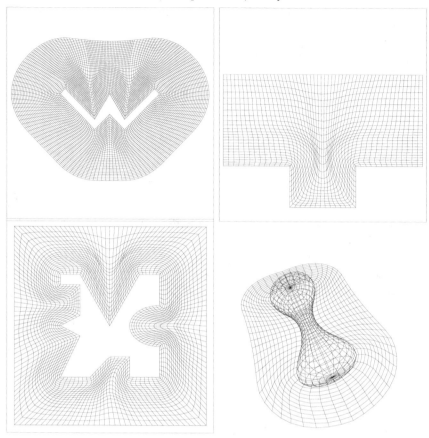

Fig. 15.5. Body-fitted grids generated using level set approach: IV.

16

Image Enhancement and Noise Removal

Outline: *The previous two sections concerned geometrical motion of a particular hypersurface of interest. In this chapter, we present schemes for image enhancement and noise removal in which all the level sets have meaning. An image is interpreted as a collection of iso-intensity contours which can be evolved. Flow under curvature both removes spikes of noise (since they correspond to high curvature objects) and smooths out oscillations in boundaries, all the while essentially preserving sharp boundaries. Variations on curvature flow lead to effective schemes for a variety of image processing applications.*

16.1 Statement of problem

Define an *image* to be an intensity map I given at each point (x, y) of a two-dimensional domain. The range of the function $I(x, y)$ depends on the type of image. For black-and-white images, the range is either 0 or 255 (here we employ the standard convention that 0 corresponds to black and 255 corresponds to white). For gray-scale images, $I(x, y)$ maps each pixel into a value between 0 and 255. For color images, $I(x, y)$ is a vector-valued function into some color-space such as RGB or HSI.

Given such an image, common aims are to remove noise from the image (see Figure 16.1) without sacrificing useful detail and to enhance or highlight certain features. This is a subjective goal; the classification of some information as "noise" and other information as "useful" detail is in the eye of the beholder, and techniques to correspondingly filter images must, at some level, reflect this subjective decision. One natural filter exploits the idea of scale – one can try to filter out information that occupies a small amount of the domain, such as spots of non-matching colors or small oscillations in boundaries of objects.

200

Fig. 16.1. An image $I(x, y)$ with noise.

One of the most straightforward and widely used methods for removing noise is the Gaussian filter, in which both one-dimensional and two-dimensional signals are smoothed by convolving them with a Gaussian kernel; the degree of blurring is controlled by the characteristic width of the Gaussian filter. To understand how this works, imagine a gray-scale intensity function $I(x, y)$, and suppose "noise" is added to this image by replacing 10% of the pixels with new values drawn randomly between 0 and 255 (see again Figure 16.1). When the intensity function is viewed as a surface plotted above the x-y plane, the changed pixels will appear as sharp upwards and downwards spikes. If this surface is convolved with a Gaussian filter, the spikes will be reduced and will blend into the background values. In this sense, the Gaussian filter removes noise. However, the Gaussian is an isotropic operator. Because it smooths in all directions, sharp boundaries will also be blurred (see Figure 16.2).

Consequently, one goal is to improve upon this basic idea and try to remove noise without too much blurring. Various techniques have been introduced to improve upon this basic idea, including Wiener filters [82], anisotropic diffusion schemes that perform intraregion smoothing in preference to interregion smoothing (see Perona and Malik [194]), and more recently wavelet processing (Ruskai et al. [210]). We now discuss some aspects of applying the level set methodology to this problem.

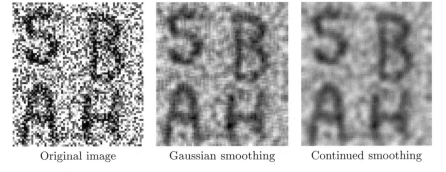

Original image   Gaussian smoothing   Continued smoothing

Fig. 16.2. Noise removal through Gaussian smoothing/heat operator.

16.2 Equations of motion

16.2.1 Curvature-driven noise reduction

As a starting point, imagine evolving the intensity $I(x, y)$ under the flow

$$I_t = \nabla^2 I. \tag{16.1}$$

This is just the heat operator applied to the image intensity, and it is easy to believe that the "noise" will be removed. However, the cost will be smearing of the sharp boundaries in the real image. First, let's rewrite Eqn. 16.1 in a slightly odd form, namely,

$$I_t = F|\nabla I| \quad \text{where} \quad F = \frac{\nabla \cdot \nabla I}{|\nabla I|}. \tag{16.2}$$

This is just our level set equation with a bizarre "speed" function and the level set function ϕ replaced by the intensity function. Note, however, that this is a corrupt view, since nothing is propagating normal to the level curves. Instead, smoothing isotropically takes place in all directions. Sharp boundaries are smeared.

Now, consider an alteration to Eqn. 16.2, which was introduced by Alvarez, Lions, and Morel [9]. If the denominator gradient is passed *inside* the divergence operator, it yields the evolution equation

$$I_t = F|\nabla I|, \quad \text{where} \quad F = \nabla \cdot \frac{\nabla I}{|\nabla I|} = \kappa, \tag{16.3}$$

where here we have noted that $\nabla \cdot \frac{\nabla I}{|\nabla I|} = \kappa$. This is our standard curvature evolution equation. The attractive quality of this approach is that sharp boundaries are preserved; smoothing takes place inside a region, but not across region boundaries.

An alternative method takes a total variation approach to the problem (see Rudin, Osher, and Fatemi [208]), and leads, once again, to a level set methodology and reduces to a very similar curvature-based speed function given by $F(\kappa) = \kappa/|\nabla I|$. Following these works, variations on these approaches were produced by Sapiro and Tannenbaum [215]; in that work, a speed function of the form $F(\kappa) = \kappa^{1/3}$ was employed.

In each of these schemes, all information is eventually removed through continued application of the scheme. This is because Grayson's theorem says that each contour shrinks to zero and disappears. Consequently, a "stopping criterion" is required. A level set scheme for noise removal and image enchancement that does not require such a stopping criterion was introduced by Malladi and Sethian [159]. The scheme results from returning to the original ideas of curvature flow, and exploiting a "Min/Max" function that correctly selects the optimal motion to remove noise. It has two highly desirable features:

(i) It defines an intrinsic, adjustable definition of scale within the algorithm. All noise below that level is removed, and all features above that level are preserved.

(ii) The algorithm stops automatically once the sub-scale noise is removed. Continued application of the scheme produces no change.

These two features are quite powerful and lead to a series of open questions about the morphology of shape and asymptotics of scale removal. In the next section, we describe this Min/Max scheme for noise removal and image enhancement; for details, see [161, 159, 162].

16.2.2 The Min/Max flow

Our starting point is the standard speed function $F(\kappa) = -\kappa$. In the rest of this discussion, we need to be particularly careful about signs. Imagine a curve initialized so that the interior of the disk has a negative value for the signed distance function ϕ and the exterior has a positive value for the signed distance function ϕ. Then the normal $\nabla\phi/|\nabla\phi|$ points outward, and the curvature defined as $\nabla \cdot \frac{\nabla\phi}{|\nabla\phi|}$ is always positive on all the convex level contours. Thus, a flow under speed function $F = -\kappa$ corresponds to the collapsing curvature flow, since the boundary moves in the direction of its normal with negative speed, and hence moves

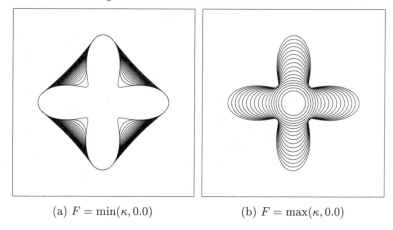

(a) $F = \min(\kappa, 0.0)$ (b) $F = \max(\kappa, 0.0)$

Fig. 16.3. Motion of curve under Min/Max flow.

inwards. We will also refer to a new speed function $\bar{F}$ in the context of a rewritten level set equation

$$\phi_t = \bar{F}|\nabla\phi|; \qquad (16.4)$$

From now on, $\bar{F} = -F$. The reason for doing so is that it produces fewer minus signs in the discussion below. Thus, a curve collapsing under its curvature will correspond to speed $\bar{F} = \kappa$. This will be the convention for the remainder of this chapter.

Now, consider two variations on the basic curvature flow given by

- $\bar{F}(\kappa) = \min(\kappa, 0.0)$
- $\bar{F}(\kappa) = \max(\kappa, 0.0)$

As shown in Figure 16.3, flow under $\bar{F}(\kappa) = \min(\kappa, 0.0)$ allows the inward concave fingers to grow outward, while suppressing the motion of the outward convex regions. Thus, the motion halts as soon as the convex hull is obtained.[1] Conversely, flow under $\bar{F}(\kappa) = \max(\kappa, 0.0)$ allows the outward regions to grow inward while suppressing the motion of the inward concave regions. Once the shape becomes fully convex, the curvature is always positive and the flow becomes the same as regular curvature flow, in which case the shape collapses to a point.

---

[1] The limiting convex hull is a consequence of our level set embedding perspective. If we just look at a single curve propagating with speed $\min(\kappa, 0.0)$, there are cases in which the curve gets "hung up" and cannot perform the topological change required to produce the convex hull. However, a level set flow under this speed function will change topology and produce the convex hull.

To summarize the above, flow under $\bar{F} = \min(\kappa, 0.0)$ preserves some of the structure of the curve, while flow under $\bar{F} = \max(\kappa, 0.0)$ diffuses away all of the information.

16.2.2.1 Min/Max flow on structures of a prescribed scale

The goal is to select the correct choice of flow that both smooths out small oscillations, and maintains the essential properties of the shape. In order to do so, we introduce the idea of the Min/Max switch. Consider the following speed function [161, 159]:

$$
\bar{F}^{\text{Stencil=k}}_{\min / \max} = \begin{cases} \min(\kappa, 0) & \text{if } \text{Ave}^{R=kh}_{\phi(x,y)} < 0 \\ \max(\kappa, 0) & \text{if } \text{Ave}^{R=kh}_{\phi(x,y)} \geq 0 \end{cases} . \qquad (16.5)
$$

where $\text{Ave}^{R=kh}_{\phi(x,y)}$ is defined as the average value of ϕ in a disk of radius $R = kh$ centered around the point (x, y).[2] Here, h is the grid size of the grid. Thus, given a "stencil radius" k, the above yields a speed function that depends on the average value of ϕ in the neighborhood of a given size, and the value of the curvature of the level curve going through (x, y).

In order to examine this speed function in some detail, consider a black region on a white background, chosen so that the interior has a negative value of ϕ and the exterior a positive value of ϕ.

- Stencil radius $k = 0$

  If the radius $R = 0$ ($k = 0$), then the choice of $\min(\kappa, 0)$ or $\max(\kappa, 0)$ depends only on the value of ϕ. All the level curves in the black region will attempt to form their convex hull, when seen from the black side, and all the level curves in the white region will attempt to form *their* convex hull. The net effect will be no motion of the zero level set itself, and the boundary will not move.

- Stencil radius $k = 1$

  If the average is taken over a stencil of radius h, then some movement of the zero level corresponding to the boundary is possible. If there are some oscillations in the front boundary on the order of one or two pixels, then the average value of ϕ at the point (x, y) can have a different sign than the value at (x, y) itself. In this case, the flow will act as if it were selected from the "other side". Some motion will be allowed until these first-order oscillations are removed and a balance

---

[2] When $k = 0$, we choose the value of ϕ at the grid point itself, rather than the limiting value.

between the two sides is again reached. Once this balance is reached, further motion is suppressed.

- Stencil radius k

    By taking averages over larger stencils, larger amounts of smoothing are applied to the boundary. In other words, decisions about where features belong are based on larger and larger perspectives. Once features on the order of size k are removed from the boundary, balance is reached and the flow stops automatically. At one extreme, let $k = \infty$. Since the average will compute to a value close to the background color, on this scale all structures are insignificant, and the max flow will be chosen everywhere, forcing the boundary to disappear.

To illustrate this hierarchical flow, we start with an initial shape in Figure 16.4(a) and first perform the Min/Max flow until a steady state is reached with stencil size zero in Figure 16.4(b); with this stencil size, the final and initial states are the same. Min/Max flow is then performed until a steady state is achieved with stencil size $k = 1$ in Figure 16.4(c), and then Min/Max flow is again applied with a larger stencil until a steady state is achieved in Figure 16.4(d).

The curve stops because of three effects. First, the curve is embedded in a level set framework. Second, the calculation is performed on a grid. Third, the "piling up" of level sets, as discussed above, causes shear in the level set function. The definition of "stop" is that the curve motion has essentially gone to zero, and that running the calculation for an extremely long time would be required for any appreciable motion to occur.

These results can be summarized as follows:

- The Min/Max flow switch selects the correct motion to diffuse the small-scale pixel notches into the boundary.

- The larger, global properties of the shape are maintained.

- Furthermore, and equally important, the flow stops once these notches are diffused into the main structure.

- Edge definition is maintained, and, in some global sense, the area inside the boundary is preserved.

- The noise removal capabilities of the Min/Max flow are scale-dependent and can be hierarchically adjusted.

- The scheme requires only a nearest neighbor stencil evaluation.

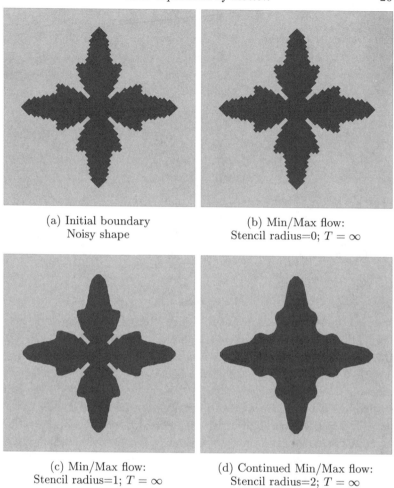

(a) Initial boundary
Noisy shape

(b) Min/Max flow:
Stencil radius=0; $T = \infty$

(c) Min/Max flow:
Stencil radius=1; $T = \infty$

(d) Continued Min/Max flow:
Stencil radius=2; $T = \infty$

Fig. 16.4. Motion of star-shaped region with noise under Min/Max flow at various stencil levels.

16.2.2.2 Extension of Min/Max scheme to gray-scale, texture, and color images

This technique applies to black and white images. An extension to gray-scale images can be made by replacing the fixed threshold test value of 0 with a value that depends on the local neighborhood. As designed in [159], let $T_{\text{threshold}}$ be the average value of the intensity obtained in the direction perpendicular to the gradient direction. Note that since the direction perpendicular to the gradient is tangent to the iso-intensity contour through (x, y), either the two points used to compute $T_{\text{threshold}}$ are either in the same region or the point (x, y) is an inflection point, in which case the curvature is in fact zero and the Min/Max flow will always yield zero. Here, choosing a larger stencil means computing this tangential average over end points located farther apart.

Formally then, a Min/Max scheme applicable to many types of images, becomes

$$\bar{F}_{Min/Max} = \begin{cases} \max(\kappa, 0) & \text{if } \text{Ave}_{\phi(x,y)}^{R=kh} < T_{\text{threshold}} \\ \min(\kappa, 0) & \text{otherwise.} \end{cases} \tag{16.6}$$

Further details about this scheme applied to a wide range of images, including salt-and-pepper and multiplicative images, and Gaussian noise applied to black and white, gray scale, textured, and color images, may be found in [161].

16.3 Results

First, we show some examples of binary images with gray-scale noise under Min/Max flow, taken from Malladi and Sethian [161, 159]. The Min/Max function switch is taken around 128, which is halfway between 0 for black and 255 for white. Figure 16.5 shows a handwritten character with noise. 10% noise means that at 10% of the pixels, the given value is replaced with a number chosen from a uniform distribution between 0 and 255. The left column gives the original figure with the corresponding percentage of noise; the right column gives reconstructed values. The figures on the right are converged; continued application of the scheme yields no significant change in the results.

Next, gray-scale noise is removed from a gray-scale image. Noise is added to the figure by replacing a prescribed percentage of the pixels with a new value, chosen from a uniform random distribution between 0 and 255. The results are obtained as follows. Begin with two levels of noise: 25% noise in Figure 16.6(a) and 50% noise in Figure 16.6(d). First, the Min/Max flow from Eqn. 16.6 is applied until a steady-state is reached in each case (Figure 16.6(b) and Figure 16.6(e)). This removes most of the noise. Then the scheme is continued with a larger threshold stencil for the threshold to remove further noise (Figure 16.6(c) and Figure 16.6(f)). For the larger stencil, we compute the average Ave over a larger disk and compute the threshold value $T_{\text{threshold}}$ using a correspondingly longer tangent vector.

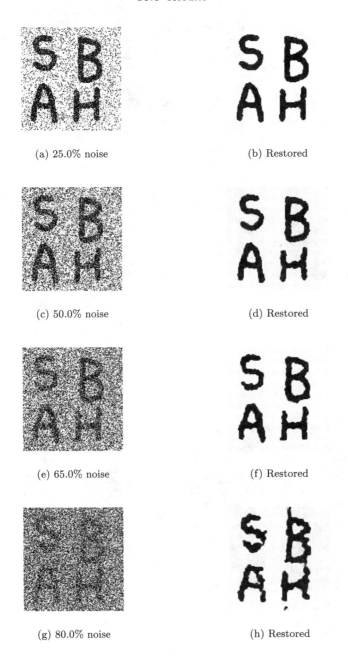

(a) 25.0% noise

(b) Restored

(c) 50.0% noise

(d) Restored

(e) 65.0% noise

(f) Restored

(g) 80.0% noise

(h) Restored

Fig. 16.5. Image restoration of binary images with gray-scale salt-and-pepper noise using Min/Max flow: restored shapes have converged ($T = \infty$).

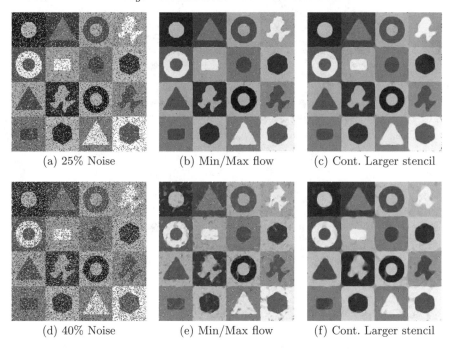

| (a) 25% Noise | (b) Min/Max flow | (c) Cont. Larger stencil |

| (d) 40% Noise | (e) Min/Max flow | (f) Cont. Larger stencil |

Fig. 16.6. Min/Max flow. The left column is the original with noise, the center column is the steady-state of Min/Max flow, the right column is the continuation to steady-state of the Min/Max flow using a larger stencil.

(a) Image with noise   (b) Min/Max: final

Fig. 16.7. Min/Max flow applied to multiplicative noise.

Next, the effect of this Min/Max scheme on multiplicative noise added to a gray-scale image is considered. Figure 16.7 shows the reconstruction of an image with 15% multiplicative noise.

These schemes can also be used to remove noise and accentuate features in medical images. In Figure 16.8, original and reconstructed images are shown. Here, no noise is artificially added; instead, the goal is to enhance certain features within the given images for later recovery.

Finally, this Min/Max algorithm can be applied to an image to which 100% Gaussian gray-scale noise has been superimposed; a random component drawn from a Gaussian distribution with mean zero is added to each (every) pixel. Figure 16.9 shows the "noisy" original together with the reconstructed Min/Max flow image. We note that the printing process itself "de-noises" images and refer the reader to a large collection of images located at http://math.lbl.gov/~malladi for a better appreciation of the noisy and reconstructed images.

16.4 Related work

A large body of work has developed around the use of curvature flows and level set methods to remove noise from images. Much of the work has concerned better operators that maintain sharp edges, anisotropic diffusion to follow texture and oriented gradients, and extensions of the techniques to color images. We refer the reader to work by Blomgren and Chan on extending total variation methods to vector-valued images [29], by Carmona on building higher order adaptive smoothing respecting feature directions [48], by Caselles, Morel, and Sbert on building fundamental axioms for image processing [51], by Chan and Wong on recovering images and blind convolution using total variation [53], by Cottet and El Ayyadi on time-delay anisotropic diffusion schemes using Volterra models [70], by Guichard on morphological, affine, and Galilean invariants and scale-spaces for movies [108], by Perona on building diffusion schemes to pick up orientations [193], by Sapiro and Tannenbaum on affine invariant scale space [214]. by Sochen, Kimmel, and Malladi on general frameworks for image diffusion and the mathematics underlying linked partial differential equation schemes for color vision processing [250, 132], by Teboul, Blanc-Féraud, Aubert, and Barlaud on variational methods for edge preserving flows [264], and by Weickert, ter Haar Romeny, and Viergever [279] on semi-implicit schemes and relieving time step restrictions for non-linear diffusion filters. For further references, we refer the reader to the book by ter Haar Romeny [206].

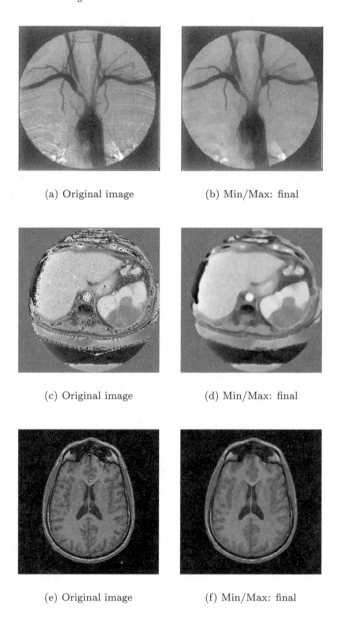

(a) Original image   (b) Min/Max: final

(c) Original image   (d) Min/Max: final

(e) Original image   (f) Min/Max: final

Fig. 16.8. Min/Max flow with selective smoothing.

(a) Original image with Gaussian noise

(b) Reconstructed Min/Max flow

Fig. 16.9. Continuous Gaussian noise added to image.

17

Computer Vision:
Shape Detection and Recognition

Outline: *In this chapter, we discuss three different aspects of computer vision: shape-from-shading, shape segmentation and recovery from images, and shape recognition. Given an image, one goal in shape detection/recovery is to isolate and build a coherent mathematical description of a particular shape that can be used in a variety of forms. The goal in shape recognition is to identify this extracted shape. Both problems can be pursued in the Level Set/Fast Marching framework.*

In this chapter, we continue our analysis of images from the perspective of what is known as "computer vision." We are interested in an entire chain of events which starts with a raw image, usually obtained from physical devices, such as computer-aided-tomography (CT) scans, and ends with an extracted shape which is then analyzed and understood.

As an example, suppose one is given a medical scan, with the goal of isolating and identifying tumors. If the tumors are manifested as areas of contrasts markedly different from the background, one might try to isolate these shapes from the image, build a mathematical/geometric model to describe the shapes, and then analyze the extracted shapes. This chained process requires several steps. First, one may want to remove noise from the original image. Next, enchancement or sharpening of particular regions can help prepare them for the segmentation step that finds the boundary of the shape. After a particular shape is segmented, recovered from the image, and then represented in a mathematical form (such as a tessellated surface or implicitly defined function), a next step may be to compute volume, surface area, geometric characteristics, etc., leading to a final stage of recognition and classification. The chain of these events is shown in Figure 17.1.

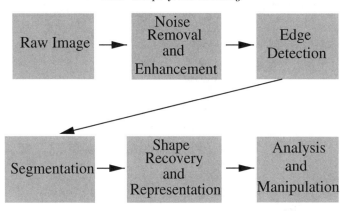

Fig. 17.1. Various stages in analysis of image data.

In the previous chapter, we have shown how Level Set Methods can be used to perform some partial differential equations-based aspects of image de-noising and enhancement. In this section, we apply both Level Set and Fast Marching Methods to aspects of edge detection, shape recovery, representation, and recognition.

17.1 Shape-from-shading

We begin with a well-known and easy-to-state problem from computer vision. This will be our first example of shape extraction from an image. The goal is to reconstruct a three-dimensional object from a single gray-scale image. To illustrate, imagine a non-self-shadowing surface, illuminated from a single (and far away) point light source. At each point of the surface, one can define the brightness map I which depends on the reflectivity of the surface and the angle between the incoming light ray and the surface normal. Points of the surface where the normal is parallel to the incoming beam are brightest; those where the normal is almost orthogonal are darkest (again, self-shadowing surfaces are prohibited). The goal of *shape-from-shading* is to reconstruct the surface from its brightness function I; one of the earliest works on this topic is provided by Horn and Brooks [115].[1]

We point out right away that the problem as posed does not have a unique solution. For example, imagine a beam coming straight up. It

---

[1] As an example, a photograph of the surface of the moon is a good candidate for a shape-from-shading reconstruction; the sun is a faraway and single light source.

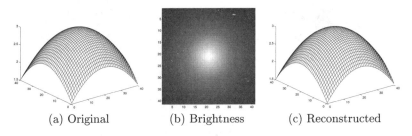

(a) Original   (b) Brightness   (c) Reconstructed

Fig. 17.2. Shape-from-shading reconstruction of paraboloid surface.

is impossible to differentiate a surface from its mirror image from the brightness function. That is, a deep valley could also be a mountain peak. Other ambiguities can exist; we refer the reader to Rouy and Tourin[207] and Kimmel and Bruckstein [130] and the recent work of Ulich [273]. Nonetheless, in its simplest form, the shape-from-shading problem provides an example of an Eikonal equation that can be solved using the Fast Marching Method.

Begin by considering an actual surface $T(x, y)$. The surface normal is then given by

$$n = \frac{(-T_x, -T_y, 1)}{(|\nabla T|^2 + 1)^{1/2}}. \tag{17.1}$$

Let (α, β, γ) be the direction from the light source. In the simplest case of a Lambertian surface, the brightness map is given by

$$I(x, y) = (\alpha, \beta, \gamma) \cdot n. \tag{17.2}$$

Thus, the shape-from-shading problem is to reconstruct the surface $T(x, y)$ given the brightness map $I(x, y)$.

We first consider the case in which the light comes straight down; that is, the viewer and the light source are located along the same direction. Then the light source vector is $(0, 0, 1)$, and thus

$$I(x, y) = \frac{1}{(|\nabla T| + 1)^{1/2}}. \tag{17.3}$$

Rearranging terms produces an Eikonal equation for the surface, namely,

$$|\nabla T| = \sqrt{\frac{1}{I^2} - 1}. \tag{17.4}$$

As initial conditions for this problem, imagine that the values of T are known at extrema of the surface. We can construct a viable solution surface using the Fast Marching Method.

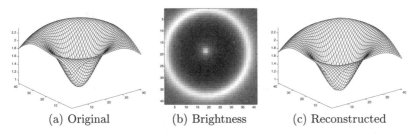

(a) Original   (b) Brightness   (c) Reconstructed

Fig. 17.3. Shape-from-shading reconstruction of double Gaussian surface.

17.1.1 Examples of shape-from-shading

To demonstrate, we start with a given surface, first compute the brightness map I, and then reconstruct the surface by solving the above Eikonal equation. Figure 17.2 shows a paraboloid surface of the form $T = 3 - 3(x^2 + y^2)$. Figure 17.2(a) shows the original surface, Figure 17.2(b) shows the brightness map $I(x, y)$, and Figure 17.2(c) shows the reconstructed surface. This surface is "built" by setting $T = 3$ at the point where the maximum is obtained, and then solving the Eikonal equation.

As a more complex example, consider a double Gaussian function of the form

$$T(x, y) = 3e^{-(x^2+y^2)} - 2e^{-20((x-.05)^2+(y-.05)^2)}. \tag{17.5}$$

Figure 17.3 shows the original figure, the brightness map and the reconstructed surface.

In the above two examples, we imagined a case when the light source was in the same direction as the viewer. The Fast Marching Method can be used in the case of an oblique light source, in which the light source vector is different from that of the viewer (see [136]). The idea is to transform to the coordinate system of the light source, so that one gets, in this new coordinate system, the Eikonal equation. To make this transformation requires some information about the surface itself. In more detail, again consider a light source vector (α, β, γ). There is an extra degree of freedom in the coordinate system; we can choose a coordinate system such that $\beta = 0$, and thus rotate the image so that $(\alpha, 0, \gamma) \cdot (0, 1, 0) = 0$. In this case, the shading image is given by

$$I(x, y) = (\alpha, 0, \gamma) \cdot n, \tag{17.6}$$

where n is defined as before. The presence of the α term means that T_x factors into the expression. Define the new coordinate system by $(\tilde{x}, \tilde{z})$

(y remains unchanged). If the brightness map $\tilde{I}(\tilde{x})$ was given in this coordinate system, then this would revert to an Eikonal equation in this new coordinate system, and the Fast Marching Method could be used directly. However, we must modify the brightness map to account for this transformation, and hence

$$\tilde{I}(\tilde{x}, y) = I(\gamma\tilde{x} + \alpha\tilde{z}, y),$$

using this reverse transformation. Thus, we may propagate outward, using the smallest value of $\tilde{z}$ to evaluate this brightness map. This then allows us to produce a upwind approximation that can be solved using the Fast Marching Method. For details, see [136].

17.2 Shape detection/recovery

We now return to the more difficult problem of edge detection, segmentation, and shape recovery. Reconsider a medical image, in which the goal is to isolate a tumor from a background image. We again imagine that the tumor corresponds to a region whose pixels are of a different grayness intensity; this may result from the particular scanning device, use of contrast agents, etc. The goal is to locate this shape within the image, and then perform shape analysis and perhaps recognition.

17.2.1 Thresholding

At first blush, the most straightforward approach to this problem is to use some sort of threshold; simply decide that the boundary of the desired shape corresponds to pixels whose values lie between the values on the inside and those on the exterior. Indeed, for regions with very sharply delimited contrasts between the interior and the exterior, this works well. In many cases, however, the raw image contains noise in the form of errant pixels whose values stray quite far from those of their neighbors. Simple thresholding may interpret small rings around these pixels as boundaries, yielding a noisy and poor representation of the desired boundary. As illustration, in Figure 17.4a, we show a white region against a gray background; a few spots of gray "noise" are shown on the white background. If we attempt to recover the boundary between the white and gray by simply displaying pixel values between the two, we get the multiple boundaries outlined in black (Fig. 17.4b).

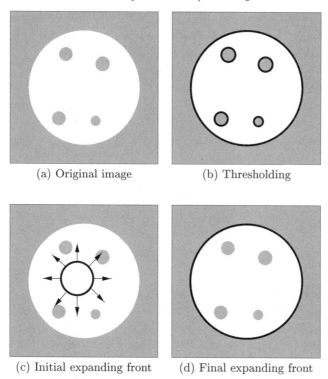

(a) Original image   (b) Thresholding

(c) Initial expanding front   (d) Final expanding front

Fig. 17.4. Thresholding vs. front propagation.

17.2.2 An expanding interface view

As an alternative, imagine propagating a front to "lock on" to the boundary. Starting from a seed (Figure 17.4c), the idea is to grow a front until it stops at the boundary (Figure 17.4d). The strategy of using level set methods for shape recovery was invented by Malladi; his idea was to exploit these techniques to grow "around" the false noise boundaries, and settle only on the larger, true boundary. There are two key components of this technique: first, the ability of our algorithms to naturally execute topological change, and second, a stopping criterion built from the image gradient.

Here, we follow the discussion in Malladi, Sethian, and Vemuri [168]; further work using the level set scheme in the context of shape recovery may be found in [158, 160, 163, 165, 166, 169] and the work of Caselles, Catte, Coll, and Dibos [49]. We refer the interested reader to those papers for motivation, details, and a large number of examples.

Given an image with the goal of isolating a shape within the image, the approach is motivated by the active force contour/snake approach to shape recovery given by Kass, Witkin, and Terzopoulos [123]. Consider a speed function of the form $\pm 1 - \epsilon\kappa$, where ϵ is a constant. As seen earlier, the constant acts as an advection term. The uniform expansion (contraction) with speed 1 (-1), depending on the sign, corresponds to the inflation force used by Cohen [68]. The diffusive second term $\epsilon\kappa$ smooths out the high curvature regions and has the same regularizing effect as the internal deformation energy term in thin-plate-membrane splines [123].

The goal now is to define a speed function from the image data that acts as a halting criterion for this speed function. Multiply the above speed function by the term

$$g_I(x,y) = \frac{1}{1+\mid \nabla\left(G_\sigma * I(x,y)\right) \mid}, \tag{17.7}$$

where the expression $G_\sigma * I$ denotes the image convolved with a Gaussian smoothing filter whose characteristic width is σ. The term $\nabla\left(G_\sigma * I(x,y)\right)$ is essentially zero except where the image gradient changes rapidly, in which case the value becomes large. Thus, the filter $g_I(x,y)$ is close to unity away from boundaries, and drops to zero near sharp changes in the image gradient. These changes presumably correspond to the edges of the desired shape. In other words, the filter function anticipates steep drops in the image gradient, and retards the evolving front from passing out of the desired region.

Thus, the algorithm works as follows. A small front (typically a single seed point) is initialized inside the desired region, grows outwards, and stops at the shape boundary as the filter term reduces the speed function F to near zero. This approach has several desirable aspects.

- The initial front can consist of many fronts; because of the topological capabilities of the level set method, these fronts will merge into a single front as it grows into the particular shape.
- The front can follow intricate twists and turns in the desired boundary.
- The technique can be used to extract three-dimensional shapes as well by initializing in a ball inside the desired region.
- Small isolated spots of noise where the image gradient changes substantially are ignored; the front propagates *around* these points, breaks into two, and the ring around the isolated spot closes back in on itself and then disappears.

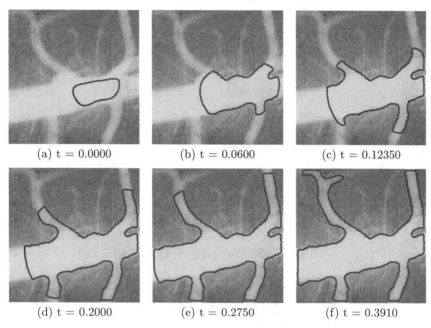

(a) t = 0.0000   (b) t = 0.0600   (c) t = 0.12350

(d) t = 0.2000   (e) t = 0.2750   (f) t = 0.3910

Fig. 17.5. Reconstruction of a shape with significant protrusions: an arterial tree structure.

As a demonstration, the recovery of the structure of an arterial tree is considered (taken from [168]). The real image has been obtained by clipping a portion of a digital subtraction angiogram. This is an example of a shape with extended branches and significant protrusions. The front is initialized in Figure 17.5(a). In subsequent frames the front evolves into the branches and finally in Figure 17.5(f) completely reconstructs the complex tree structure. Thus, a single initialization of the shape model sprouts branches and recovers all the connected components of a given shape. Calculations were carried out on a 64×64 grid with a time step of $\Delta t = 0.001$.

17.2.3 Combining algorithms

Two additional tools yield a complete approach. First, the shape recovery work can be pre-processed with the Min/Max approach to noise removal and enhancement. By doing so, artificial boundaries in the image, which may slow down or derail the evolving front, can be removed. Second, the Fast Marching Method can be used to greatly accelerate the initial propagation from the seed structure to the near boundary; Narrow Band Level Set Methods are then used to achieve accurate final results. This is the approach taken in [167] (see also [162]), and is now the preferred technique.

In more detail, we consider flow evolution of the form

$$\phi_t + g_I(1 - \epsilon\kappa)|\nabla\phi| - \beta\nabla P \cdot \nabla\phi = 0. \qquad (17.8)$$

This equation consists of three terms:

- A driving expansion (or deflation) force which is synthesized from the image gradient. Here, we take this speed function to be of the form

$$F_{\text{expand}}(x) = g_I(x) = \frac{1}{1+ |\nabla G_\sigma * I(x) |}.$$

- A surface tension force which depends on the curvature and acts as a sort of elastica on the front, namely

$$F_{\text{curv}}(x) = -g_I(x) - \epsilon\kappa.$$

This term and the previous one form the model introduced by Malladi, Sethian, and Vemuri [168] and used extensively in [158, 160, 163, 165, 166, 169].

- A force which attracts the surface towards the boundary, which has a stabilizing effect, especially when there is a large variation in the image gradient value. This term denotes the projection of an (attractive) force vector on the surface normal. This force, introduced in [50], is realized as the gradient of a potential field. Here,

$$P(x) = -|\nabla (G_\sigma * I(x)) |$$

attracts the surface to the edges in the image; the coefficient β controls the strength of this attraction.

The two-stage approach is as follows. First, we solve for the flow using the main driving term and the Fast Marching Method. Thus, we solve

$$g_I|\nabla T| = 1$$

until the solution has almost reached the desired edge; this occurs where the solution $|\nabla T|$ is large, and hence the flow has almost stopped. This provides initial conditions for the Narrow Band Level Set Method, and the full edge-detection/segmentation problem (Eqn. 17.8) is then begun. This is the approach introduced in [162].

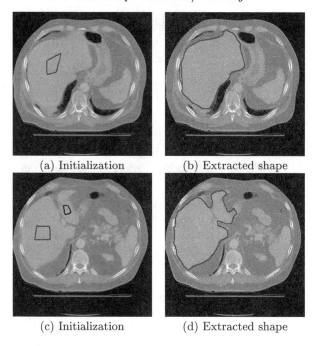

(a) Initialization   (b) Extracted shape

(c) Initialization   (d) Extracted shape

Fig. 17.6. Shape extraction of liver data.

In Figure 17.6, taken from Malladi and Sethian [162], these combined techniques are applied to the reconstruction of a liver from two-dimensional slices. Results of two different initialization sequences are shown. In Figure 17.6(a), a single contour is used in the initialization. In Figure 17.6(b) the final shape is obtained. Figure 17.6(c) shows an initialization with two separate contours of different height slices; here, the two fronts merge, and in Figure 17.6(d) the final shape is obtained.

Next, level set shape recovery techniques are applied to the difficult problem of extracting the left and right ventricles of the heart. In these calculations, the problem is initialized by simultaneously tagging both the left and right ventricles; both are found at the end of stage one by the evolving fronts. Note that in the evolution of the right ventricle front, the papillary muscle is also found (see Figure 17.7); this results from a single contour that wraps around the papillary muscle and separates into an inner and outer ring. After the inner walls of the left and right ventricles are recovered, the outer wall of the right ventricle is extracted. This is done by temporarily relaxing the stopping criterion, and allowing the front to move past the inner wall of the right ventricle. Once this

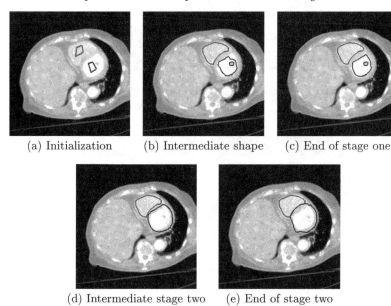

(a) Initialization   (b) Intermediate shape   (c) End of stage one

(d) Intermediate stage two   (e) End of stage two

Fig. 17.7. Shape extraction from heart data.

occurs, the stopping criterion is turned back on, and the front expands in stage two until the outer wall is found.

This technique for shape detection/recovery can be performed in three dimensions if three-dimensional data are available. Figure 17.8 shows a fully three-dimensional reconstruction of two femurs and surrounding thighs, taken from [160].

We show a few additional three-dimensional reconstructions to indicate the power of this approach. In Figure 17.9, we show a three-dimensional reconstruction of the liver (the z axis from head to toe is not to scale). Figure 17.9a shows a two-dimensional slice and the outlined two-dimensional contour; Figure 17.9b shows the full three-dimensional reconstructed shape displayed embedded in the same two-dimensional slice.

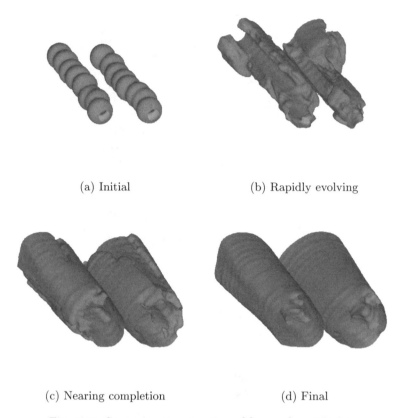

(a) Initial

(b) Rapidly evolving

(c) Nearing completion

(d) Final

Fig. 17.8. Stages in reconstruction of femurs from 3D data.

Finally, in Figure 17.10, we show the final snapshot of a full recon-struction of the brain from MRI data. The results are obtained by start-ing with some spheres as initial data and employing the Fast Marching Method with suitable stopping criterion, followed by the Narrow Band Level Set Method to provide an accurate final portrait. We refer the reader to additional work using our Level Set/Fast Marching technology for shape segmentation by Paragios and Deriche [189] and Kichenas-samy, Kumar, Olver, Tannenbaum and Yezzi [127], as well as a partic-ularly detailed application to three-dimensional confocal image analysis in cytology by Sarti, Ortiz, Lockett and Malladi [216].

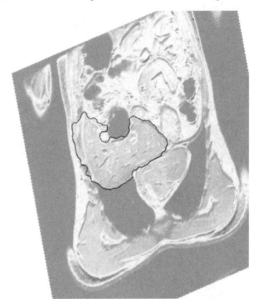

(a) Single two-dimensional contour

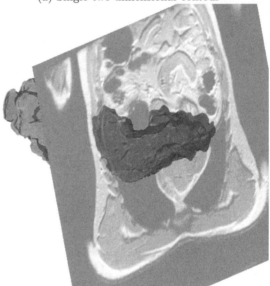

(b) Three-dimensional reconstruction

Fig. 17.9. Reconstruction of three-dimensional liver.

Fig. 17.10. Reconstruction of cortical structure.

17.3 Surface evolution and the stereo problem

An algorithm for solving the stereo problem has recently been developed by Faugeras and Keriven [95] using many of the ideas of surface evolution according to Level Set Methods and the shape segmentation approach just described. Here, we briefly describe that work; for a clear exposition and details, see the original work [95].

The stereo problem may be thought of as follows. Several two-dimensional images of a three-dimensional scene are recorded; the goal is to use these two-dimensional images to reconstruct the three dimensional scene. Given that the orientation and placement of the cameras which record the two-dimensional scenes are known, the goal is to establish correspondence between a particular pixel in one image and its location in the other frames; this is known as "registration". The idea is illustrated in Figure 17.11, taken after the one in [95].

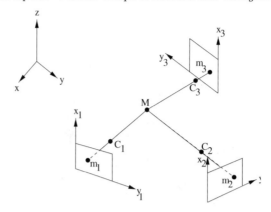

Fig. 17.11. The stereo problem: given a pixel M_1 in image 1, the goal is to find the corresponding pixel M_i in image i such that all of these pixels correspond to the original three-dimensional point M.

Once these pixels are found for every point in all the images, the rays may be traced backward to construct the three-dimensional image.

Faugeras and Keriven treat this problem as follows. First, they view the problem as a variational problem involving a functional which must be minimized in order to obtain the best registration of the images with each other. This then leads to Euler-Lagrange equations for this functional, yielding a set of partial differential equations which are interpreted as the flow of a surface toward the solution, using some of the above segmentation and level set flow ideas.

The mathematical derivation of this approach is somewhat lengthy. We shall give only a brief sketch of the argument. Faugeras and Keriven build a level set function u from R^3 to R; the coordinates of the points in the scene which are on the surfaces of the objects present are defined by the zero level set $u(x, y, z) = 0$. The goal is to start from an initial surface S and flow it in its normal direction under the flow field

$$S_t = \beta N,$$

where the function β is related to the functional which measures the matching between the images. The idea is to drive this flow under the functional being minimized. We will write down neither the exact form of the functional nor the actual evolution equation; suffice it to say that the equation includes as its first term the driving term used in the previous shape segmentation algorithm. The other terms are particular to the stereo problem. The resulting algorithms are appealing, and we refer the reader to [95] for the complete details and test examples.

17.4 Reconstruction of obstacles in inverse problems

The next application could rightfully be included in the later chapter on optimization and first arrivals, since it has little to do with computer vision. But the techniques of shape segmentation and recovery are the fundamental tools at play, and hence we include it here.

Consider a problem in which the unknown is a given region which is either simply or multiply connected. As data, one has a known input and the output response, as well as an understanding of the process by which the given region affects the output. For example, a typical inverse scattering problem is one in which the response to an unknown object is measured under a given input, and the goal is to locate the object itself. One can view the previous sections on medical imaging, segmentation and shape recovery as an inverse problem; the scanning device is the input, the output is the image density pixel map, and the unknown is the position and shape of the object that created the pixel map. In a series of clever papers, Santosa [213] and Litman, Lesselier, and Santosa [148] use a level set approach to find the unknown region by constructing a speed function F whose motion seeks to minimize the residual of a given functional. Here, we briefly discuss the work by Santosa; for details, see [213].

As an example, and following [213], imagine the inverse scattering problem in which we are given a differential equation A such that

$$A(u) = g \qquad (17.9)$$

and g is the data of the problem. The goal is to find the region D such that

$$u(x) = \left\{ \begin{array}{c} u_{\text{int}} \text{ for } x \text{ in } D \\ u_{\text{ext}} \text{ for } x \text{ not in } D \end{array} \right\}. \qquad (17.10)$$

In one example, u_{ext} is the sound speed of the exterior propagating medium and u_{int} is replaced by the boundary conditions on the boundary ∂D. Define $W(u)$ as

$$W(u) = \frac{1}{2} \|A(u) - g\|_2^2. \qquad (17.11)$$

Thus, the goal is to construct a value for the solution u which minimizes W in the least squares sense. The first step in [213] is to characterize

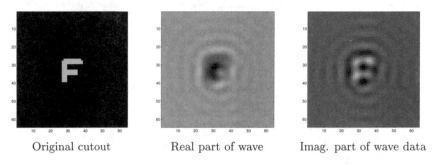

| Original cutout | Real part of wave | Imag. part of wave data |

Fig. 17.12. Exact solution and received diffraction wave (from Santosa, 1996).

the boundary using a level set function ϕ that is negative inside D and positive outside. Thus, the boundary (which is as of yet unknown) is the zero level set of ϕ. The goal is to evolve this level set function in such a way that the final answer minimizes this functional. After some manipulation, Santosa then produces a level set evolution equation of the form

$$\phi_t + [J(u)^T(A(u) - g)]|\nabla\phi| = 0, \qquad (17.12)$$

where $J(u)^T$ is the transpose of the Jacobian of $A(u)$ at u (see [213] for the derivation). Thus, the problem has been converted into a familiar level set form, with the following features: $W(u)$ is non-decreasing and the flow stops (that is, $\phi_t = 0$) when the actual solution $A(u) = g$ is achieved.

Santosa produces two nice examples using this approach, first of deconvolution in two dimensions, and second of reconstruction of a diffraction screen. We reproduce his results of the second to give some idea of the approach. The problem concerns a diffraction screen reconstruction. An opaque screen with a cutout of unknown shape is subjected to a harmonic plane wave with wave number k propagating normal to the cutout. The real and imaginary values of the plane wave are recorded downstream, and the goal is to reconstruct the cutout domain. In Figure 17.12, taken from [213], the original cutout and the received real and imaginary parts of the wave are shown.

In Figure 17.13, the evolution of the zero level set under the speed given in Eqn. 17.12 is shown; the cutout is nicely captured by the evolving front. We refer the reader to the original paper [213] for details

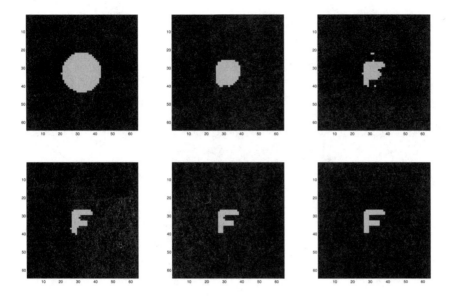

Fig. 17.13. Evolution of zero level set to extract cutout shape (from Santosa, 1996).

and further examples, as well as to the work of Litman, Lesselier and Santosa [148].

17.5 Shape recognition

In this section, we discuss two projects to couple Level Set/Fast Marching segmentation with recognition. The ultimate goal, starting with a raw image, is to extract and identify particular regions located in complex images.

17.5.1 Pre-processing recognition for shape reconstruction

Given a satellite image of the earth, one goal is to process this image to categorize meteorological data. The image may contain several different types of objects, including sea, land, shallow convective clouds, and deep convective clouds. Many current automated techniques use versions of neural net classifiers which classify individual pixels or small regions, using as input particular spectral and texture features, occurring in both visible and infrared channels. Such classifiers respond with

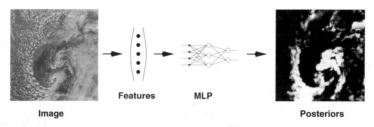

Fig. 17.14. Classification of raw image using neural net (MLP) classifier.

posterior probabilities for each pixel, indicating the likelihood of that pixel belonging to one of the categories. Thus, the output is a set of pixel intensities. An appealing next step is to segment these pixel images to connect the boundaries and extract a shape model. This is what was done in [38]; here, we review that work as an example of shape recognition coupled with our front evolution approach. We refer the interested reader to [38] as well as [37] for considerably more material and background.

Starting with a meteorological image, as shown in Figure 17.14, a set of feature vectors is chosen, and then a neural net classifier is built to yield posterior probabilities. The output of these probabilities is shown as a gray-scale pixel map on the far right.

In order to perform a Level Set/Fast Marching segmentation of this image, we must synthesize an appropriate speed function from the image output. Since there are multiple pixel densities at each pixel location due to the various posterior probabilities, it is not straightforward to produce a single speed law F. A clever way of blending these probabilities to produce an appropriate speed law is given in [37]; we refer the reader there for details. Briefly, the various posterior probabilities are combined, together with weights, to produce a best fit.

Applying the Level Set/Fast Marching framework for image segmentation and shape recovery is now straightforward, using the methodology described earlier in this chapter. Figure 17.15 shows shape recovery of the deep convective posterior probability image for two choices of a smoothing parameter ϵ which controls the amount of curvature used in the speed function.

Finally, in Figure 17.16, we show the shape extraction results for two different cloud types, overlaid on top of the original images. These images are taken from [38]; we refer the reader to the original work and to [37] for additional results and a more detailed discussion.

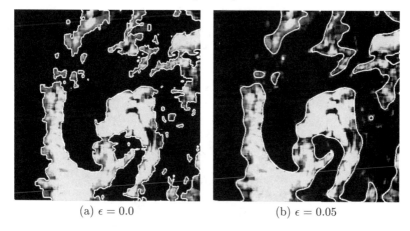

(a) $\epsilon = 0.0$ (b) $\epsilon = 0.05$

Fig. 17.15. Shape recovery of a deep convective cloud for two different smoothing parameter values.

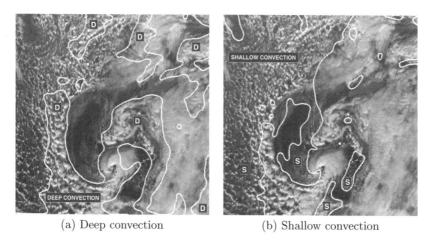

(a) Deep convection   (b) Shallow convection

Fig. 17.16. Shape recovery of deep convection and shallow convection overlaid onto original visible image.

17.5.2 Handwritten character recognition

Given a shape that has been represented in some fashion, suppose we try to use a neural net classifier to help identify the shape. This is a notoriously hard problem, touching on psychology, artificial intelligence, and cognitive and computer science. One of the first steps is to find a representation of the shape boundary that is amenable to comparison with other shapes. Two common representations are:

- *Boundary-based approaches:* Here, the goal is to focus on the boundary of the shape. In these techniques, chain codes, introduced by Freeman [96], encode a shape by traversing the shape boundary assuming either a 4 or an 8 connectivity. Polygonal approximations (Pavlidis [191]) describe a complex boundary by a sequence of straight lines. Shape features such as one-dimensional moments (Gonzalez and Wintz [104]) and Fourier descriptors (Pearson and Fu [192]) are computed from the shape boundary. In addition, these features can be made invariant to similarity and affine transformations (Arbter et. al. [15] and Crimmins [75]). Such an invariant representation markedly aids in recognition techniques.

- *Region-based approaches:* Here, the goal is to focus on the region as a whole. Among the region-based shape representation schemes, one of the most typical is a "skeleton" approach (Blum [30]) or a medial-axis transform approach (Lee [146]). There are many skeletonization algorithms; see Leymarie and Levine [151], Mayya and Rajan [172], Ogniewicz and Ilg [184], and references therein. Although skeletons provide a compact shape description, they may contain a large number of redundant edges that may minimally contribute to shape information. Efficient edge pruning schemes are discussed in [139, 172, 184].

A different approach is to use a level set perspective to represent and recognize shapes. Once again, the boundary of the shape is represented as the zero level set of the signed distance function; we note that other distance transforms have been used before for image analysis, see Danielson [76] and Borgefors [31]. There are several advantages to this approach. First, boundary- and region-based representations are implicitly contained in the signed distance function; the boundary is simply its zero level set. Second, the notion of "holes" in the boundary is seamlessly contained in this representation; unlike a boundary-based representation, a small "hole" in the object does not create an entire new set of features in the object. Third, and most importantly, a level set representation permits the evolution of the shape which can be used to accentuate certain features. One possibility is to exploit the level set evolution equation both to classify and to "perturb" boundaries to help find a way to classify them.

As an example, consider the shape of a handwritten numeral 2 in Figure 17.17(a). This shape can be represented by its boundary or by its skeleton, as shown in Figure 17.17(b). Alternatively, the shape boundary can be embedded in the signed distance function ψ defined in a square

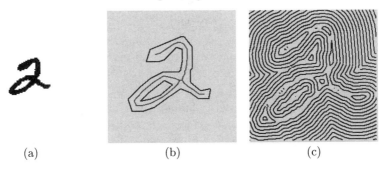

(a)   (b)   (c)

Fig. 17.17. Representations of the numeral 2

region enclosing it. Figure 17.17(c) shows the level sets (equal distance contours) of the function ϕ. Furthermore, a level set representation can be modified to find an acceptable classification.

This was the approach taken by Malladi and Sethian [158], where a level set perspective was used to classify handwritten characters (optical character recognition). To be sure, there exists a full gamut of shape features for this problem, and a thorough review on historical development of OCR (optical character recognition) technology may be found in Mori et al. [180]. Such schemes include statistical classification methods with global feature analysis schemes such as moments and mathematical transforms (Fourier, Walsh, wavelet), and syntactical methods used with structural features such as loops, junctions, strokes, convexities, etc. (see Suen et al. [257]). By no means are we suggesting that level set methods offer an optimal way to perform this task. However, they offer an interesting way to explore the notion of shape perturbation, and we now describe this approach in somewhat more detail.

17.5.2.1 Training the neural network

To begin, a database was built from a large collection of numeral samples extracted from the NIST special databases 3 and 7 of segmented characters. First, the signed distance function was computed on a square grid for a collection of numeral shapes. A back-propagation neural network classifier was then used as a recognition system. The network was trained on a random set of N sample shapes. The network weights are stored and then tested against N testing samples. The feature vector was the signed distance function expressed on a 16×16 grid. Briefly, the neural network was a fully-connected architecture consisting of 256 input nodes, 10 internal nodes, and 10 output nodes corresponding to 10 numeral classes.

A series of preprocessing steps must be performed prior to training. First, a noise removal step was executed; second, the boundary of each numeral was traced using a chain code procedure. Following this, size normalization was achieved by finding the bounding rectangle of every numeral shape and mapping it into a square region of side 0.5 and centered at the origin. During this processing step, care must be taken not to skew the shape artificially either horizontally or vertically. Next, the signed distance function was computed in the square region by considering all the contours found by the chain code procedure.

Given the level set signed-distance function description of character shapes, a multiple neural net approach is used to achieve high reliability as follows. The different neural nets correspond to using different underlying grids to evaluate the signed distance function. This can be thought of as a version of adaptive mesh refinement: a feature vector results from a particular underlying grid. The network is first trained on one feature vector grid using a random set of 3189 numeral samples and upon convergence is tested on another set of 3158 samples. Given a testing numeral sample, the trained network classifies it with a confidence value lying in the interval $[0 \dots 1]$. Reliability is improved by accepting only classifications made with high confidence and discarding the rest. The neural net classification is repeated using a different feature vector obtained by evaluation on a different grid; results are accepted only if both classification networks agree. The best results obtained using this approach yield a rate of 99.56% correct, with 12.88% discards, on NIST database 3, and 99.18% correct, with 38.65% discards, on NIST database 7 (database 7 is considerably harder). Here, "discards" are characters that do not fall into acceptance criteria levels, and hence the network does not offer a guess as to their nature.

While the results are encouraging, given the small number of characters used in the training, the idea of using successive sieves of network classifiers means that relatively large numbers of discards are generated. These discards are then subjected to "perturbation."

17.5.2.2 Perturbing the characters: searching neural net space

The strategy is to evolve the signed distance function under the level set equation to generate perturbations of characters in the discard set. These perturbations are constantly checked against the existing neural net classifiers for acceptance. If one of the perturbations moves within the classification range, it is accepted. Hence, perturbations are performed in neural net space, creating an expanding ball around the given discard, and the first point on that ball that falls into some network's acceptance criteria is accepted.

What remains is to generate a set of flow rules that form the perturbations. Flow rules taken from previous level set applications include:

- *Curvature flow:* The interface (that is, the boundary between the inside and outside of a character) is moved inward in its normal direction according to its curvature. As shown previously, this flow smooths out oscillations in the shape boundary.

- *Outward constant flow:* The interface propagates outward with constant speed (see [225, 228]); as discussed previously, this flow removes oscillations and closes off holes.

- *Horizontal anisotropic flow:* As discussed in [243], anisotropic flow is an integral part of crystal growth and dendritic solidification. The anisotropic symmetry gives preferred directions of growth. Twofold symmetry in the horizontal direction is taken, which means that shapes move under speed laws that prefer to move them in horizontal directions.

- *Vertical anisotropic flow:* The same as the above, except that flow in the vertical direction is preferred.

- *Growing limbs:* One would like a flow that performs the opposite of the curvature flow, in that limbs (regions of high positive curvature) grow outward. However, as is clear from [225], this is an ill-posed (and numerically unstable) backward heat equation. Instead, a scheme was invented which grows limbs for a short time. Begin by flowing under curvature flow for a given number of time steps. Then, using this final shape and the initial shape, interpolate backwards in time to a previous state, which will have larger "limbs" at places where the initial curvature is positive. Repeat the process of curvature flow to smooth oscillations in the front, plus backward time interpolation to extend limbs farther. This process is unstable, in that refining the time step produces uncontrollable oscillations. However, done on a coarse enough scale, it produces shapes that exaggerate and accentuate limbs and protrusions.

As an example, in Figure 17.18, the perturbation of a sample hand-written numeral 9 is shown under several different flow rules.

Using these flow rules, the discarded characters are perturbed and

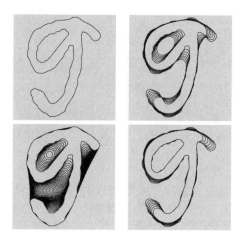

Fig. 17.18. Perturbations of numeral under flow rules.

then compared with the neural net classifiers. Under this strategy, recognition rates of 99.08%, with 6% discard, are obtained for NIST database 3; for NIST database 7, 98.05% correct is achieved with around 20% discard. This represents a considerable improvement over the static results. A sample of the characters that this technique successfully recognizes is shown in Figure 17.19.

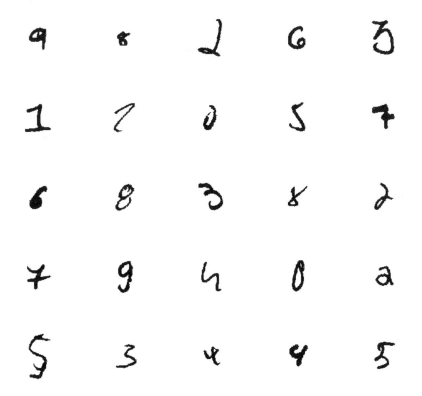

Fig. 17.19. Hand-written NIST characters correctly identified.

18

Interface Methods for Combustion, Solidification, Fluid Mechanics, and Electromigration

Outline: *In this chapter, we review some applications of Level Set Methods to interface problems in which the front acts as an internal boundary condition to a partial differential equation and the solution of this equation controls the motion of the front. In combustion problems, the interface is a sharp reaction zone which both drives the fluid dynamics and is driven by the velocity field. In solidification problems, the interface represents a solid/liquid boundary, and jump conditions across the interface drive the dynamics. In fluid mechanics problems, the interface separates fluids, and the role of surface tension and/or density differences between the two fluids are important. In electromigration problems, moving internal boundaries provide conditions for elliptic and parabolic differential equations. Such problems fit nicely into the extension velocity framework and algorithms presented in Chapter 11.*

In this chapter, we review some applications of Level Set Methods to interface problems in which the front acts as an internal boundary condition to a partial differential equation, and the solution of this equation controls the motion of the front. Such problems fit nicely into the extension velocity framework and algorithms presented in Chapter 11, in which the velocity at the front itself, obtained by solving a set of partial differential equations on either side of the interface (together with some jump conditions), is then extended to the neighboring level sets, and then all the ϕ grid values are updated within the narrow band.

As examples, we consider problems from combustion, solidification, fluid mechanics, and electromigration, as well as some other applications. In combustion problems, the interface is a sharp reaction zone which both drives the fluid dynamics and is driven by the velocity field. In solidification problems, the interface represents a solid/liquid boundary, and jump conditions across the interface drive the dynamics. In fluid mechanics problems, the interface typically separates fluids of different densities and/or viscosities, and the role of gravity and/or surface tension are important. In electromigration problems, moving internal boundaries provide conditions for elliptic and parabolic differential equations.

The common link in these applications is the presence of a term in the equations of motion represented by a Dirac delta function along the interface. This term will induce the interplay between the interface and the partial differential equations off of the front. Our emphasis will be on the algorithmic issues involved in applying the level set formulation to the setting under discussion, rather than a detailed discussion of the relevant physics. We shall try to give some sense of how these level set applications evolved, as well as touch on the results. Some of these applications do not employ the extension velocity methodology of Chapter 11. In these cases, we will try discuss the basic ideas used to construct a velocity field for the neighboring level sets from the partial differential equation itself. The interested reader is referred to the literature cited below for detailed descriptions of the algorithms and results.

18.1 Combustion

In many combustion phenomena, it is reasonable to view a propagating reaction zone as an infinitely thin flame front. In this view, the combustion reaction dynamics happen over a thin zone relative to the underlying fluid mechanics; resolving the width and inner structure of the flame may not be as important as modeling the larger-scale interactions between the flame and fluid mechanics. This view of a flame as an infinitely thin zone lends itself to a level set formulation, in which the flame is a sharp interface propagating normal to itself with a speed function F. The interactions between the flame motion and the fluid dynamics can be quite intricate and depend on such factors as the following:

- **The burning velocity:** The flame speed in the normal direction, relative to the underlying fluid velocity, can depend on temperature, pressure, and the fuel mass fraction which represents the species in the system; this flame speed can be modified by the effects of curvature, following a model due to Markstein [170].

- **Exothermic expansion:** Along the flame front, the fluid can undergo a change in density corresponding to expansion as it is converted from reactants to products; thus, the flame can act as a delta function source of volume.

- **Vorticity production:** Pressure gradients tangential to the flame cause different accelerations in light and heavy fluids. This causes a production of vorticity across the flame, since the pressure gradient is not always aligned with the density gradient.

- **Underlying fluid mechanics:** Fluid dynamics transports the flame, causing mixing and bringing new reactants into contact with the flame/high temperature zone; this in turn modifies the burning and hence creates a velocity field which influences the hydrodynamics.

There are many additional factors that go into this model, which is known as premixed turbulent combustion. As one might guess, the various scales at work require delicate and careful schemes. The interplay among flame zones, acoustic waves, large pressure variations and temperature gradients is subtle.

The implementation of level set methods in this context is quite natural. The flame is viewed as a zero level set of a higher-dimensional function, and the effects of curvature and normal direction are easily computed from the level set function. Here, we discuss in some detail three different uses of level set methods for these combustion problems. One is based on a Lagrangian vortex method for tracking the fluid mechanics, another relies on a finite difference scheme for solving Euler equations in conservation form, and the third deals purely with the geometry of detonation shock dynamics.

18.1.1 *Vorticity, exothermicity, flame stretch, and wrinkling*

In a study by Rhee, Talbot, and Sethian [202], a flame is viewed as an infinitely thin reaction zone separating two regions of different but constant densities. The hydrodynamic flow field is two-dimensional and inviscid, and the Mach number is vanishingly small. This corresponds to the equations of zero Mach number combustion, introduced by Majda and Sethian [156]. The flame propagates into the unburnt gas at a prescribed flame speed S_u that depends on the local curvature.

The reason for this flame speed dependence on curvature comes from the role of heat conduction. Imagine a slightly perturbed planar flame front. The concave part of the perturbed flame is surrounded by burnt gases, which provide more heating to the reactants than would occur with a purely planar interface. Conversely, the convex part of the perturbed flame is surrounded by unburnt gases, which heat up the reactants less. This model, due to Markstein [170], indicates that the flame speed depends on the local curvature, that is,

$$S_{\text{unburnt}} = S^o_{\text{unburnt}}(1 - L\kappa), \tag{18.1}$$

where S^o_{unburnt} is the speed of a planar flame and L is a constant. For details on this model, see [202, 224, 100, 223].

The flame is represented as the zero level set whose speed is controlled by local curvature and whose position is advanced according to both burning and hydrodynamic advection under the fluid flow field. At the same time, the position of the interface acts as a source of exothermic volume on the right-hand side of a Poisson equation for the velocity. In addition, the local stretch of the interface in the tangential direction acts as a source of vorticity that also contributes to the hydrodynamic flow field. We now explain this model in more detail and follow the discussion in [202].

18.1.1.1 Equations of motion

Following the derivation by Pindera and Talbot [196], the velocity U is decomposed into the three components

$$U = U_s + U_v + U_p, \tag{18.2}$$

where U_s is the incompressible velocity field due to exothermic expansion along the front, U_v is the rotational velocity field due to stretch-induced vorticity, and U_p is the potential velocity of the incident flow. Thus, the individual component fields satisfy

$$\nabla \cdot U_s = m \ \delta(x - x_f); \nabla \times U_s = 0, \tag{18.3}$$

$$\nabla \times U_v = \omega(x); \nabla \cdot U_v = 0, \tag{18.4}$$

$$U_p = \nabla\Phi; \nabla \cdot U_p = 0, \tag{18.5}$$

where m is the volume source strength per unit length associated with the amount of expansion along the flame front, x_f is the location of the flame, $\omega(x)$ is the vorticity field, Φ is the vector potential of the incident flow, and $\delta(x)$ is the two-dimensional Dirac delta function. The volume expansion m is related to the normal velocity S_{unburnt} of the flame on the unburnt side through

$$m = \frac{\rho_{\text{unburnt}} - \rho_{\text{burnt}}}{\rho_{\text{burnt}}} S_{\text{unburnt}}, \qquad (18.6)$$

where ρ_{burnt} (ρ_{unburnt}) is the density in the burnt (unburnt) gases. The vorticity transport equation, obtained by taking the curl of the Navier–Stokes equation (discussed in the next section) is given by

$$\frac{D}{Dt}\left(\frac{\omega}{\rho}\right) = \left(\frac{\omega}{\rho}\right)\nabla \cdot U + \nu\nabla^2\omega + \frac{1}{\rho^2}\nabla\rho \times \nabla P, \qquad (18.7)$$

where P is the pressure, ρ is the density, and ν is the viscosity.

A formula for the vorticity jump $[\omega]$ was produced by Hayes [111] in the inviscid limit, and is given by

$$[\omega] = \left(\frac{1}{\rho_{\text{burnt}}} - \frac{1}{\rho_{\text{unburnt}}}\right)\nabla_\tau\left(\rho_{\text{unburnt}}S_{\text{unburnt}}\right) \qquad (18.8)$$

$$- \frac{\rho_{\text{burnt}} - \rho_{\text{unburnt}}}{\rho_{\text{unburnt}}\rho_{\text{burnt}}}\left[\frac{dU_\tau}{dt} + U_\tau(\nabla_\tau U_\tau - V_n\kappa) - V_n\frac{\partial V_n}{\partial\tau}\right],$$

where τ is the tangential direction, U_τ is the flow velocity at the flame in the direction tangential to the flame, ∇_τ is its gradient along the flame, and V_n is the absolute normal flame speed. Here, d/dt is the time derivative taken at a point which lies on the front. Defining the flame stretch K as

$$K = \frac{1}{A}\frac{dA}{dt} = \nabla_\tau U_\tau - V_n\kappa, \qquad (18.9)$$

where A is an elemental flame front area, we then have

$$[\omega] = \left(\frac{1}{\rho_{\text{burnt}}} - \frac{1}{\rho_{\text{unburnt}}}\right)\nabla_\tau\left(\rho_{\text{unburnt}}S_{\text{unburnt}}\right) \qquad (18.10)$$

$$- \frac{\rho_{\text{burnt}} - \rho_{\text{unburnt}}}{\rho_{\text{unburnt}}S_{\text{unburnt}}}\left[\frac{dU_\tau}{dt} + U_\tau K - V_n\frac{\partial V_n}{\partial\tau}\right]. \qquad (18.11)$$

This gives the amount of vorticity produced due to flame propagation.

18.1.1.2 Algorithm and construction of extension velocities

The flame is tracked by identifying the flame interface as the zero level set of the level set function. The curvature is determined using the expression given in Eqn. 6.35. The vortical field U_v is represented by a collection of vortex blobs using Chorin's vortex method [64]; see also

[229]. The exothermic field is determined by solving a Poisson equation on the underlying grid with the given right-hand side by smearing the Dirac delta function to the neighboring grid points. The no-flow boundary is satisfied by the addition of a potential flow that exactly cancels the existing flow field. Finally, the tangential stretch component is evaluated by tracing back the values of U_τ along a normal vector from the given position backward and normal to the front to evaluate the derivative $V_n = \frac{\partial U_\tau}{dt}$.

The algorithm explicitly marches ahead in time as follows. Imagine that the various fluid velocities U_v, U_s, and U_p are all known at a given time step, as is the level set function ϕ that describes the location of the flame. The algorithm advances from one time step to the next through the following steps:

(i) Given the level set function, the local curvature is evaluated.

(ii) Cells containing the front are located, and volume sources m are extrapolated to nearby grid points. If Δl is an element of flame, then the strength $m = \left(\frac{\rho_{\text{unburnt}} - \rho_{\text{burnt}}}{\rho_{\text{burnt}}} \right) S_u \Delta l$, is area-weighted to neighboring grid points.

(iii) The volumetric flow field is determined by solving the Poisson equation $\nabla^2 \Psi = f_{ij}$, where f_{ij} are the volume weights. Then we have $U_s = \nabla \Psi$.

(iv) The flame stretch is computed, and new vortex elements are added to the flow by evaluating the right-hand side of Eqn. 18.11.

(v) The vortical flow U_v is calculated by using a vortex method over the total accumulated vortices.

(vi) The entire velocity field is assembled.

(vii) The potential flow U_p is computed to account for the incident flow field. A suitable Neumann boundary condition is supplied so that the total flow satisfies the no-slip boundary condition.

(viii) The vortex elements are updated, and the level set function is advanced under both curvature-controlled burning and hydrodynamic advection.

In terms of extension velocities, we note the following:

• Flame speed: The flame velocity F in the normal direction is determined from Eqn. 18.1. This is evaluated for all the level sets, not just the zero level set, and this provides a natural velocity for all the level sets.

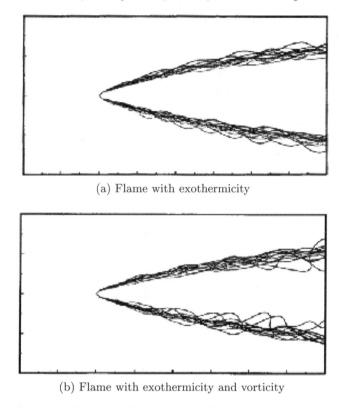

(a) Flame with exothermicity

(b) Flame with exothermicity and vorticity

Fig. 18.1. Comparison of flame brush.

- The exothermic velocity U_s: The Poisson equation is solved on a finite difference grid, and the computed velocity at each grid point serves as the extension velocity.

- The vorticity velocity U_v: The Biot-Savart law is used to compute the associated velocity field everywhere from the vortex elements; this produces an extension velocity at the grid points.

- The boundary conditions U_p: This too is solved on a fixed grid using a fast matrix solver and provides the extension velocity field everywhere.

Because the extension methodology of Chapter 11 was not used in [202], periodic re-initialization of the level set function was required. This was performed by finding the front and recomputing the exact signed distance.

18.1.1.3 Results

Figure 18.1, taken from [202], shows two results from this algorithm used to model an anchored flame, with upstream turbulence imposed by a statistical distribution of positive and negative vortices. The goal is to understand the effects of exothermicity and flame-induced vorticity on the flame wrinkling and stability. In Figure 18.1(a), an anchored flame in the oncoming turbulent field under the effects of exothermic expansion due to the density jump across the flame is shown; here, different time snapshots are superimposed upon each other to show the flame "brush." In Figure 18.1(b), the effects of both volume expansion and vorticity generation along the flame front are included. The resulting flow field generates a significantly wider flame brush, as the vorticity induces flame wrinkling which influences the exothermicity of the surrounding flow field.

18.1.2 A hybrid scheme for tracking deflagrations

The technique of building a numerical scheme to explicitly locate and describe an interface is known as "front tracking"; conversely, an algorithm that solves for a physical variable everywhere and characterizes the interface by certain values of this variable (or where its gradient is sharp) is known as "front capturing". In other words, front tracking techniques explicitly represent the front (such as with marker particles), while front capturing techniques locate the interface using a variable defined everywhere. Smiljanovski, Moser, and Klein [247] have devised a hybrid front capturing and tracking algorithm for deflagration combustion fronts using a level set method. Here, we briefly discuss that work; for details, see [247].

In [247], the Euler equations are solved on a fixed mesh using high order upwind finite difference schemes for hyperbolic conservation laws. Their approach is applicable to problems in which strong density jumps (shocks) interact with the combustion dynamics. The coupling between the fluid dynamics and the flame is as follows. A fixed regular grid is employed. The level set function is used to explicitly locate the front, calculate the normal to the front, and extract volume fractions of pre- and post-discontinuity states. The underlying Euler equations are then solved throughout the domain, incorporating the effects of temperature and density. Unlike front tracking methods, which treat the two sides of the flame front separately, and hence suffer from oddly shaped regions when the front cuts through a cell at awkward places, accurate numerical fluxes for the whole cell are built in [247], using detailed information about the local geometry provided by the level set method. Knowing the front geometry, separate pre- and post-front numerical flux densities are obtained, then yielding updated values.

The central idea is thus to track the front using a Level Set Method, and then incorporate local geometric information to construct appropriate flux averages in the cell. By doing so, one can treat general propagation with all the advantages of the level set view of front tracking, such as topological change and the ability to calculate geometric quantities, while still maintaining the attractiveness of a scheme that executes in essentially the same manner across all cells.

The velocity field produced from the Euler solution is used to provide the front velocity for the neighboring level sets, hence the gradient of ϕ can stretch or bunch up, requiring re-initialization; this re-initialization is performed using the iterative technique suggested in [262].

In [247], this technology is applied to the interaction of a flame with a shock in one dimension, the burn out of a combustion chamber in one dimension, two-dimensional merger of flame kernels, and an analysis of two-dimensional transition to detonation, in which flame-induced pressure waves precondition the gas and the resulting feedback leads to shock formation and, finally, detonation. We refer the reader to [247] for these and other applications.

18.1.3 Modeling detonation shock dynamics

In [16], Aslam, Bzdil, and Stewart uses the geometrical aspects of Level Set Methods to track evolution of detonation fronts. Detonation shock dynamics is an asymptotic theory that describes the evolution of a multidimensional and curved detonation shock. If one were to treat this system in its entirety, the resulting equations would include the compressible Euler equations, an equation of state, and a reaction rate law. However, an asymptotic analysis produces an intrinsic surface evolution equation for the detonation shock velocity of the form (see [16])

$$D_n = D_{\text{CJ}} - \alpha(\kappa),$$

where D_{CJ} is the one-dimensional steady Chapman-Jouget velocity for the explosive, and $\alpha(\kappa)$ is a function of the curvature κ which is a material property of the explosive. This falls naturally under the analysis of the speed law $F = 1 - \epsilon\kappa$ studied in [222, 225]. What makes this application interesting is the incorporation of angle boundary conditions which describe the interaction of detonation waves with inert boundaries at material interfaces. As the shock hits this boundary, a series of angle-related boundary conditions must be applied. Since the detonation shock (as tracked by the zero level set function) is not necessarily aligned with the material interface, Aslam, Bzdil, and Stewart employ an additional static (unchanging) level set function which characterizes the material boundary; this is similar to the sub-grid masking work described earlier in Chapter 12. This auxiliary level set function provides orientation information about the material boundary.

In their approach, a way of converting the evolving detonation shock to the appropriate one-dimensional problem normal to this interface is required. The idea is to exploit the Immersed Interface Method, due to LeVeque and Li [150]; in these techniques, coefficients in the partial differential equation are adjusted through one-sided Taylor series expansions to account for the presence of the interface. Thus, the idea is to solve the purely geometric surface evolution of detonation shocks in interior regions, which are coupled together through appropriate boundary conditions induced at material interfaces.

Using this approach, a series of problems are computed, including the response of an initially planar Chapman-Jouget detonation front to the sudden loss of confinement at a straight edge, the formation of a Mach reflection when a detonation front enters a converging channel, and the diffraction of detonation shock when entering a diverging channel. We refer the interested reader to [16] for results and further discussion.

18.2 Crystal growth and dendritic solidification

Next, we discuss an application of Level Set Methods to front propagation which links interface motion to a heat diffusion equation on either side of the evolving interface, together with some jump conditions. We discuss two approaches applying these evolution techniques to the problem of crystal growth and dendritic solidification.

18.2.1 Background

Imagine a container filled with a liquid such as water, which has been smoothly and uniformly cooled below its freezing point so that the liquid does not freeze. The system is now in a "metastable" state, where

a small disturbance such as dropping a tiny seed of the solid phase into
the liquid will initiate a rapid and unstable process known as *dendritic
solidification.* The solid phase will grow from the seed by sending out
branching fingers into the distant cooler liquid nearer the undercooled
wall. This growth process is unstable; small perturbations of the initial data can produce large changes in the time-dependent solid–liquid
boundary.

Mathematically, this phenomenon can be modeled as a moving boundary problem. The temperature field satisfies a heat equation in each
phase, coupled through two boundary conditions on the unknown moving solid–liquid boundary, as well as initial and boundary conditions on
the container walls. The moving boundary conditions explicitly involve
geometric properties of the boundary itself, such as the local curvature
and the normal direction, as well as the temperature field. For further
details, see Cahn and Hilliard [42], Gurtin [109], and Mullins and Sekerka
[182].

18.2.2 Equations of motion

Following [243], the model under consideration includes the effects of
undercooling, crystalline anisotropy, surface tension, molecular kinetics,
and initial conditions. Consider a square container B, filled with the
liquid and solid phases of some pure substance. The unknowns are the
temperature $u(x,t)$ for x in B, and the solid–liquid boundary $\Gamma(t)$.

The temperature field u is taken to satisfy the heat equation in each
phase, together with an initial condition in B and boundary conditions
on the container walls:

$$u_t = \nabla^2 u \text{ in } B \text{ off } \Gamma(t), \tag{18.12}$$

$$u(x,t) = u_0(x) \text{ in } B \text{ at } t = 0, \tag{18.13}$$

$$u(x,t) = u_B(x) \text{ for } x \text{ on } \partial B. \tag{18.14}$$

Since the position and velocity of the moving boundary $\Gamma(t)$ are unknown, two boundary conditions on $\Gamma(t)$ are required to determine u
and $\Gamma(t)$. Let n be the outward normal to the boundary, pointing from
solid to liquid. The first boundary condition is the classical Stefan condition:

$$\left[\frac{\partial u}{\partial n}\right] = -HV \text{ on } \Gamma(t). \tag{18.15}$$

Here $[\partial u/\partial n]$ is the jump in the normal component of heat flux $\partial u/\partial n$ from solid to liquid across $\Gamma(t)$, V is the normal velocity of $\Gamma(t)$, taken positive if the liquid is freezing, and the constant H is the dimensionless latent heat of solidification. The signs of geometric quantities are chosen so that if $\partial u/\partial n < 0$ in the liquid phase and $\partial u/\partial n = 0$ in the solid phase, then $[\partial u/\partial n]$ is negative and $V > 0$, indicating that the solid phase is growing. The speed V of the interface is linked to the jump in heat flux across the boundary. Physically this means that undercooling drives solid growth. The latent heat of solidification controls the balance between geometry and temperature effects.

The second boundary condition on $\Gamma(t)$ is the classical Gibbs–Thomson relation, modified to include crystalline anisotropy and molecular kinetics as well as the surface tension:

$$u(x,t) = -\epsilon_\kappa(n)\kappa - \epsilon_V(n)V \text{ for } x \text{ on } \Gamma(t). \qquad (18.16)$$

This says that the temperature on the interface depends on the surface tension and the velocity V. Here κ is the curvature at x on $\Gamma(t)$, taken positive if the center of the osculating circle lies in the solid. The anisotropy functions are modeled by

$$\epsilon_\kappa(n) = \epsilon_\kappa(1 - A\cos(k_A\theta + \theta_0)), \qquad (18.17)$$

$$\epsilon_V(n) = \epsilon_V(1 - A\cos(M_A\theta + \theta_0)), \qquad (18.18)$$

where θ is the angle between n and the x axis, and ϵ_κ, ϵ_V, A, M_A, and θ_0 are constants depending on the material and the experimental setup. For example, if $\epsilon_\kappa = 0$ ($\epsilon_V = 0$), there are no surface tension (molecular kinetic) effects. For $A = 0$, the system is isotropic, while if $A > 0$, the solid is M_A-fold symmetric with a symmetry axis at angle θ_0 to the x axis. Typically $A \leq 1$. For a sketch, see Figure 18.2.

18.2.3 A boundary integral approach

A variety of techniques can be used to approximate numerically these equations of motion. One approach is to solve the heat equation numerically in each phase and try to move the boundary so that the two boundary conditions are satisfied; see Chorin [66], Smith [249], Kelly and Ungar [125], Meyer [177], Sullivan, Lynch, and O'Neill [258], and Dziuk and Schmidt [79, 219]. Another approach is to recast the equations of motion as a single integral equation on the moving boundary

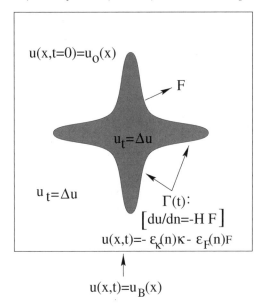

Fig. 18.2. Equations of motion and configuration.

and solve the integral equation numerically, as is done by Kessler and Levine [126], Langer [143], Meiron [175], and Strain [255].

The first application of Level Set Methods to crystal growth and dendritic solidification was by Sethian and Strain [243]. In that work, information about the speed of the interface is extended to the other level sets by means of a boundary integral along the front that is evaluated everywhere in space. The central idea is to exploit a transformation due to Strain [255], that converts the equations of motion into a single, history-dependent boundary integral equation on the solid–liquid boundary that can be evaluated by a combination of fast techniques. This boundary integral equation is given by

$$\epsilon_\kappa(n)\kappa + \epsilon_V(n)V + U + H \int_0^t \int_{\Gamma(t')} K(x, x', t - t')\, V(x', t')dx'\, dt' = 0$$

$$(18.19)$$

for all x on the interface $\Gamma(t)$, where K is the heat kernel. Note that the velocity V depends not only on the position of the front but also on its previous history. Thus, as shown by Strain [255], information about the temperature off the front is stored in the previous history of the boundary.

As in other level set applications, the interface is then identified with the zero level set of ϕ. All that remains is to extend the velocity to the grid points so that all of the level sets can be advanced. This is done by evaluating Eqn. 18.19 throughout the narrow band. An accurate evaluation of the integral then comes from breaking it up into two components: first, a history part that contains information about the past beyond a certain time, and second, a local part that is accurate in a close space/time neighborhood. The advantage to this split is that each part can be evaluated by its own accurate and fast technique (see Strain [254, 255, 256] and Greengard and Strain [107]). In order to evaluate this history integral, the zero level set is found and used to produce a set of quadrature points. These points are then used to evaluate the integral equation and compute the velocity V at each grid point.

The use of a boundary integral allows one to calculate results for an absolutely sharp interface. No smearing of the surface tension or heat release are present. As such, it allows one to compute the evolution of sharp boundaries without the diffusive effects that come from a grid. In [243], an extension velocity is constructed by evaluating the boundary integral both on and off the front. With the more modern technology of Chapter 11, a new approach would be to rely on the Fast Marching methodology to first compute this boundary integral on the front, and then use the extension velocity which preserves the distance function.

18.2.3.1 Results

Figure 18.3 shows one example from [243], in which the effect of changing the latent heat of solidification H is analyzed. Since the latent heat H controls the balance between the pure geometric effects and the solution of the history-dependent heat integral, increasing H puts more emphasis on the heat equation/jump conditions. Calculations are performed on a unit box, with a constant undercooling on the side walls of $u_B = -1$. The kinetic coefficient is $\epsilon_V = .001$, the surface tension coefficient is $\epsilon_\kappa = 0.001$, and there is no crystalline anisotropy ($M_A = 0$, $A = 0$). The initial shape was a perturbed circle with average radius $R = 0.15$ and perturbation size $P = 0.08$ and $L = 4$ limbs. A 96×96 mesh is used with time step $\Delta t = 0.00125$. The calculations are all shown at the same time.

In the calculations shown, H is varied. In Figure 18.3(a), $H = 0.75$ and the dominance of geometric motion serves to create a rapidly evolving boundary that is mostly smooth. H is increased in each successive figure, ending with $H = 1.0$ in Figure 18.3(d). As the latent heat of solid-

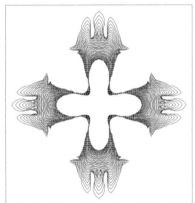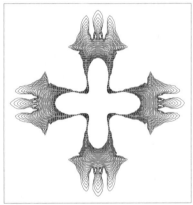

Upper left: $H = 0.75$ Upper right: $H = 0.833$
Lower left: $H = 0.916$ Lower right: $H = 1.0$

Fig. 18.3. Effect of changing latent heat.

ification is increased, the growing limbs expand outward less smoothly, and instead develop flat ends.

These flat ends are unstable and serve as precursors to tip splitting. Note that the influence of the heat integral slows down the evolving boundary. Presumably, increasing latent heat decreases the most unstable wavelength, as described by linear stability theory. The final shape shows side branching, tip splitting, and the strong effects of the side walls.

18.2.4 A full grid approach

A later application of these interface techniques was developed by Chen et. al. in [55]. In that approach, the authors work directly with the Eulerian formulation of the problem, that is, with a heat equation described in each region, using a finite difference scheme for solving the heat equation. The virtue of this approach is that the entire problem is treated on a single grid as a pure finite difference problem.

There are several points to note about this approach. A straightforward explicit finite difference scheme is used away from the interface. At and near the interface, one-sided (upwind) differencing is used, with the values of the curvature computed using the standard level set approach. The Gibbs–Thomson relation is used as a boundary condition for the construction of these one-sided operators; the velocity F of the interface is built by evaluating the temperature jump. In order to use this technique, an extension velocity is built off of the front by solving an advection equation. The chosen advection equation requires re-initialization of the front at every time step, using the methodology of [262]. This full domain approach was used to produce a collection of nice results, and we refer the reader to [262].

18.3 Fluid mechanics

There have been numerous applications of Level Set Methods to interface tracking in fluid mechanics, including problems in two-phase flow, bubble dynamics, and compressible flow. Here, we summarize some of those works.

Two early applications of fluid dynamics problems to track the interface were the projection method calculations of compressible gas dynamics of Mulder, Osher, and Sethian [181] and the combustion calculations of Zhu and Sethian [285]. Each viewed the interface as the zero level set and tracked the interface to separate the two regions.

To begin, Mulder, Osher, and Sethian [181] studied the evolution of rising bubbles in compressible gas dynamics. The level set equation for the evolving interface separating two fluids of differing densities was incorporated inside the conservation equations for the fluid dynamics. Both the Kelvin–Helmholtz instability and the Rayleigh–Taylor instability were studied; the density ratio was about 30 to 4, and both gases were treated as perfect gases. Considerable discussion was devoted to the advantages and disadvantages of embedding the level set equation as an additional conservation law. These early level set calculations suffered from considerable smoothing of the fluid variables near the interface, in part because of the particular choice of extension velocities (the fluid velocity throughout the region was used, without any attempt to either extend the correct speed from the front or re-initialize the evolving level set function), and in part because of incomplete understanding of the interactions between this formulation and the fluid solvers. These issues were addressed in later papers, discussed below.

Next, in the combustion calculations of Zhu and Sethian [285], the interface was a flame propagating from the burnt region into the unburnt region. Unlike the above calculations concerning flame stability in flame holders, in these calculations the flame was viewed as a "cold flame": the hydrodynamic flow field affected the position of the flame, but the advancing flame did not in turn affect the hydrodynamic field. In these calculations, the hydrodynamic field was computed using Chorin's projection method [63], coupled with the level set approach; in fact, a second order version developed by Bell, Colella, and Glaz [25] was used. The problem under study was the evolution of a flame inside a swirling two-dimensional chamber. The results showed the intermixing that can occur, the creation of pockets of unburnt fuel surrounded by burnt pockets, and the relationship between flame speed and mixing levels.

These two works were followed by Level Set Methods applied to the motion of incompressible, immiscible fluids in which steep gradients in density and viscosity existed across the interface. In these problems, the role of surface tension was crucial and formed an important part of the algorithm. A valuable series of calculations using a level set approach were carried out by Sussman, Smereka, and Osher [262] and Chang, Hou, Merriman, and Osher [54]. As in [285], the hydrodynamic flow field, was updated using a second order projection method.

We now discuss these ideas in more detail. Begin with the Navier–Stokes equations, that is,

$$u_t + (u \cdot \nabla)u = F + \frac{1}{\rho}(-\nabla P + \mu \nabla^2 u + \mathrm{ST}). \qquad (18.20)$$

Here, u is the fluid velocity, F is a forcing term (typically gravity), ρ is the fluid density, μ is the fluid velocity, P is the pressure, and ST is the surface tension of the interface. Assume a sharp fluid interface between two fluids with different densities ρ_1 and ρ_2, and also that the flow is incompressible, and thus

$$\nabla \cdot u = 0. \qquad (18.21)$$

The surface tension term acts normal to the fluid interface and is proportional to the curvature, due to a balance of force argument between the pressure on each side of the interface. This leads to the relation

$$\text{ST} = \sigma\kappa\delta(d)n, \tag{18.22}$$

where σ is the coefficient of surface tension, κ is the curvature, n is the normal to the interface, d is the distance to the front. and $\delta(d)$ is the Dirac delta function. Thus, surface tension acts as an additional forcing term in the direction normal to the fluid interface. Once again, similar to the combustion example, the strategy is to smear this delta function to neighboring grid points.

Using a formulation developed by Brackbill, Kothe, and Zemach [34], Sussman, Smereka, and Osher [262] and Chang, Hou, Merriman, and Osher [54] performed this smearing as follows. Begin with the surface tension expression $\sigma\kappa\delta(d)n$. By replacing the normal n by $\nabla\phi/|\nabla\phi|$, and noting that the distance d is approximated by $\phi/|\nabla\phi|$ (this is just the slope formula), we have that

$$\sigma\kappa\delta(d)n = \sigma\kappa(\phi)\delta(\phi)\nabla\phi. \tag{18.23}$$

This recasts the surface tension in the level set framework. If ϕ is always re-initialized to the distance function, the delta function itself can be smoothed over several grid points, using one of many smoothing operators. Some further details are required; in particular, one must take special care in the differencing of the curvature term necessitated by the projection method. For details, see [262].

Thus, the equations of motion become as follows:

$$u_t + (u \cdot \nabla)u = F + \frac{1}{\rho}(-\nabla P + \mu\nabla^2 u + \sigma\kappa(\phi)\delta(\phi)\nabla\phi), \tag{18.24}$$

$$\nabla \cdot u = 0. \tag{18.25}$$

These equations are solved using variations of Chorin's [63] projection method, which works as follows. The method is a fractional step method in which a temporary updated velocity field u^{new} is produced at time step $n+1$ by solving the above Navier-Stokes equation. By using the Hodge decomposition, this temporary velocity field is decomposed into two components, namely

$$u^{\text{new}} = \nabla w + v,$$

where $\nabla \cdot v = 0$; w then corresponds to the pressure, and the incompressible part of the velocity is then extracted to satisfy the requirement that $\nabla \cdot u = 0$ (for details, see [63]). The surface tension term, which acts only on the interface itself, is smeared out to the underlying finite difference grid by taking the derivative of a smoothed Heaviside function (see [262, 260]); this results in some smearing of the actual physics in the case of large density ratios between the different fluids.

In Zhu and Sethian [286], the fall of a heavy fluid bubble in a lighter density fluid was studied using these techniques as a test calculation for a fully three-dimensional code under development. Figure 18.4 shows the results of one such calculation. The initial shape is a circle, and a sequence is shown as the bubble falls. The density ratio between the heavy and light fluids is 2 to 1; in the time evolution, one sees the development of the typical spiral vortical rollups.

These basic ideas coupling between Level Set Methods to fluid mechanics have been revisited in a collection of recent papers aimed at improving the efficiency of the calculations. In [260], considerable attention is focused on constructing a re-initialization technique that does not artificially move the fluid interface every time step. Sussman and Smereka [261] studied axisymmetric three-dimensional free boundary problems using these approaches, including calculations of a rising bubble bursting at a free surface. Using the Immersed Interface Method of LeVeque and Li [150], Hou, Li, Osher and Zhao studied Hele–Shaw flow using a level set method in [116]. We refer the reader to these, as well as [54, 262, 286], for further details.

Finally, we mention the work of Zhang, Zheng, Prasad, and Hou [281], which constructs a curvilinear level set method for deformable free surface problems arising in solidification. In this work, level set methods are coupled with multizone adaptive grid generation techniques on which a curvilinear finite volume discretization is employed. As before, the underlying Navier-Stokes equations include the effects of surface tension and density variation. The adaptive grid code is based on a constrained adaptive optimization approach, and the fluid equations are updated using a finite difference approximation. Using this approach, the authors have studied mold filling and spreading, merger, and solidification of molten droplets on a cold substrate. We refer the reader to [281] for a detailed explanation and results.

18.4 Additional applications

Here, we briefly cover a few more application areas for Level Set Methods coupled to external partial differential equations.

$t = 0$

$t = 0.2$

$t = 0.4$

$t = 0.6$

$t = 0.7$

$t = 0.8$

Fig. 18.4. Falling bubble: density ratio 2:1.

18.4.1 Groundwater flow

Holm and Langtangen [114] addressed the simulation of groundwater flow for two-phase porous media, where the phases are separated by a sharp interface. They assume two incompressible, homogeneous fluids in a porous medium and assume that the saturation of each fluid is either zero or one. Thus, there is no macroscopic mixing of the two fluids. Darcy's law applies for each fluid, and the effects of permeability, viscosity, pressure, density and gravity are included. A finite element Petrov-Galerkin formulation is used, and mesh refinement about the interface is employed. Using this approach, they studied a series of problems, including the stability of a water-air system, motion of fresh and saltwater interfaces, seepage through a dam, and contamination transport. We refer the reader to [114] for further details.

18.4.2 Film boiling

Son and Dhir [252] extended level set methods for two-phase flow to include the effects of liquid-vapor phase change in a study of the mechanisms behind bubble release patterns in the evolution of the liquid-vapor interface. An energy equation for film boiling is derived to satisfy the condition that the vapor-liquid interface is maintained at the saturation temperate; this provides an additional boundary condition. The effects of volume expansion due to liquid-vapor phase change, similar to those in the combustion processes described earlier, are included in the mass conservation equation. As in the previous methods, Chorin's projection method is used to solve the time evolution of the velocity and pressure. Using these techniques, Son and Dhir perform numerical simulations and compare with the experiment the effect of wall superheat on bubble release and the change in this mechanism under increasing temperatures. For details, see [252].

18.4.3 The Marangoni effect

Chen, Bi, and Jiang [56] study the liquid bridges that form due to differential heating and surface tension in fluids. At issue is the stability of the flow surface which arises in the processing of liquid metals or other small Prandtl number fluids. The work in [56] studies the position of the free boundary that arises in internally heated circulating flows under surface tension. This extension of the level set work on fluid mechanics includes the effect of temperature as part of the evolution equations. A semi-implicit scheme is used to integrate the equation; an implicit scheme is used for the viscous terms and an explicit Adams-Bashforth scheme is used for the non-linear convection terms.

18.5 Void evolution and electromigration

We end this chapter with an application of Level Set Methods coupled with Immersed Interface Methods to study void evolution and electromigration in the study of the reliability and failure of materials. The basic ideas is to study the failure of materials resulting from the migration of vacancies under diffusion and driven by electric potentials. These vacancies grow in voids, which can eventually become large enough to cause a breakdown in the electrical properties of the metal. The work described is taken from [245], and a much more detailed explanation with many examples may be found in that reference.

18.5.1 Set up and equations of motion

The basic problem may be thought of as follows. Imagine a two-dimensional rectangular domain A which represents a metal. An electric potential $\psi(x,y)$ will be put across the domain by requiring that the potential be a fixed constant on the left wall, and that the derivative $\frac{\partial \psi}{\partial n}$ be given on the right wall. The top and bottom walls have insulating boundary conditions corresponding to no current flow through the walls, i.e., $\frac{\partial \psi}{\partial n} = 0$. Inside the metal are vacancies represented by V which diffuse according the diffusion equation and are driven by the electric potential; boundary conditions are also supplied for the vacancy concentration $V(x,y)$. Finally, as part of the initial conditions, a set of void regions denoted by Γ are supplied. Boundary conditions for the potential ψ and the vacancy concentration V are given on the void/metal boundaries. The setup and variables are shown in Figure 18.5.

The problem evolves under time as follows. The position Γ of the void boundary supplies boundary conditions for a Poisson problem for the potential. This provides a driving force for the diffusion of vacancies in the metal, which is represented through an advection-diffusion equation for the concentration. The solution of this equation provides a flux for the growth of the void boundary, and the change in the void shape and position creates a new electric potential which affects the vacancy distribution.

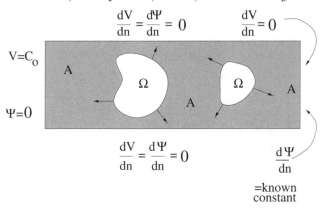

$$\frac{dV}{dn} = \frac{d\Psi}{dn} = 0 \qquad \frac{dV}{dn} = 0$$

$V = C_0$

$\Psi = 0$

$$\frac{dV}{dn} = \frac{d\Psi}{dn} = 0$$

$$\frac{d\Psi}{dn}$$

=known
constant

Fig. 18.5. Setup for electromigration problem.

We may thus write the equations of motion as follows. Let Γ be the void boundary. Our goal is to solve the following:

- **Electric potential**: Given the position Γ of the voids, find the electric potential ψ in the region A satisfying the Poisson equation:

$$\nabla^2 \psi = 0, \qquad (18.26)$$

subject to the boundary conditions

$$\frac{\partial \psi}{\partial n} = 0 \quad \text{on } \Gamma$$

$$\psi^n = 0 \text{ on left}, \quad \frac{\partial \psi}{\partial n} = \text{ Constant on right}$$

$$\frac{\partial \psi}{\partial n} = 0 \quad \text{on top and bottom}$$

- **Vacancy diffusion/transport**: Solve the advection/diffusion equation for the vacancy concentration:

$$\frac{\partial V}{\partial t} = \left[\nabla \cdot \left[D \nabla V^n - \frac{V^n}{l} \nabla \psi^n \right] \right],$$

subject to the boundary conditions

$$V(x, y, t) = C_0 \text{ on left} \quad \frac{\partial V}{\partial n} = 0 \text{ on right}$$

$$\frac{\partial V}{\partial n} = 0 \text{ on top and bottom} \quad V = 0 \text{ on } \Gamma$$

- **Void motion**: Solve the evolution of the void boundary Γ under the normal velocity field given by

$$F = \frac{\partial V^{n+1}}{\partial n}\Big|_\Gamma \qquad (18.27)$$

18.5.2 Flow of algorithm

In [245], these equations of motion are solved using a combination of Narrow Band Level Set Methods and Immersed Interface Methods (see [150]). The idea is to use an operator-split technique in which one first solves for the potential given the position of the void boundary, then uses this potential to update the vacancy equation, and then computes the vacancy flux to advance the position of the void boundary. To begin, the boundary Γ is represented by the zero level set of ϕ. Given the boundary, the potential ψ is obtained by solving Poisson's equation for the potential in the entire domain, and adjusting the five-point stencil at the boundary by using Taylor series to account for the internal boundary condition. This gives the potential in the domain A. The advection-diffusion equation is then solved with the given boundary conditions. The normal derivative of the concentration provides the normal velocity F to the interface. The extension methodology discussed earlier is used to build a velocity field throughout the narrow band, which is then used to update the level set function which advances the void boundary. The sequence is then repeated.

18.5.3 Results

Here, we briefly describe some results on these and some modified equations. For complete details, see [245]. We can consider a non-dimensional set of equations, namely,

$$\nabla^2 \psi = 0 \quad \text{Potential equation,} \qquad (18.28)$$

$$V_t = \nabla^2 V + \beta \nabla V \cdot \nabla \psi \quad \text{Concentration evolution equation,} \qquad (18.29)$$

$$F = \alpha \frac{\partial V}{\partial n}\Big|_\Gamma \quad \text{Void growth equation} \qquad (18.30)$$

We can make the assumption that the concentration diffusion is fast relative to the void growth. Thus we are led to a steady-state diffusion equation at each time step of the form

$$0 = \nabla^2 V + \beta \nabla V \cdot \nabla \psi \qquad (18.31)$$

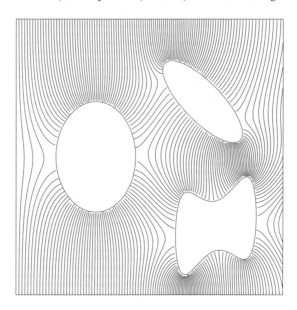

Fig. 18.6. Calculation of potential around voids.

Given this quasi-steady state solution, we can then compute the concentration flux to find the interface speed, and then update the interface.

To begin, in Figure 18.6 we show the ability of the Immersed Interface Method to compute the potential around several arbitrary regions. As expected, the potential lines are correctly calculated to be normal to the void boundaries.

Figure 18.7 shows an example of the growth of a single void on a 80×80 grid. We chose $\alpha = 0.001$ and $\beta = 2.0$. On the left, an initial circular vacancy grows into an oblong shape in response to the potential and the concentration diffusion equation. A time sequence is shown, and at the final state the accompanying equi-concentration curves are also shown. On the right, the potential function contours are shown.

Finally, we show the growth of multiple voids using this approach, again taken from [245]. Figure 18.8 shows two simulations. On the left, three voids are considered with $\beta = 0$, and on the right, with $\beta = 10.0$. Thus, the left figure ignores the effect of vacancy transport term and void growth relies only on vacancy diffusion, while the right figure includes the electric transport effect. On the left, equi-concentration lines are drawn for the last shape computed, while on the right lines of constant potential

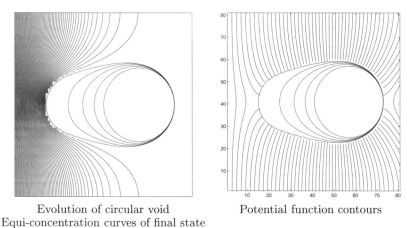

Evolution of circular void   Potential function contours
Equi-concentration curves of final state

Fig. 18.7. Growth of a single void.

are drawn. The boundary conditions for this problem are different from those given for the previous case, and are chosen so that a circle would grow radially outward so that the enclosed area increases at a constant rate.

For a detailed discussion of the coupling of Level Set Methods and Immersed Interface Methods to problems in electromigration and for additional results, see [245].

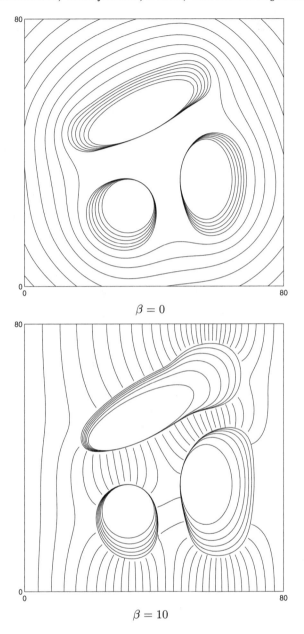

Fig. 18.8. Growth of three voids.

19

Computational Geometry and Computer-aided Design

Outline: *In this chapter, we explore the application of our techniques to problems in computational geometry and computer-aided-design (CAD). The key is to represent curves and surfaces, both as building blocks and as complex boundaries, using implicit functions. We begin with shape-offsetting, and then the use of these techniques to construct Voronoi diagrams. We then discuss techniques to construct minimal surfaces and surfaces of prescribed curvature, and end with algorithms for shape and surface smoothing.*

19.1 Shape-Offsetting

The problem of *shape-offsetting* is straightforward; given a closed shape in two or three dimensions, compute the offset, obtained by propagating the boundary in its normal direction with constant unit speed. This easy example is part of a much larger set of tools required to machine parts in an efficient and automatic fashion.

All that is required is to compute the distance to the boundary and plot the level curves. A technique for doing so using a level set approach was given by Kimmel and Bruckstein [131]. The Fast Marching Method offers an alternative and highly efficient approach by working directly with the Eikonal equation.

Since the speed is given by $F = 1$, we then must solve

$$|\nabla T| = 1, \tag{19.1}$$

where the initial condition is $T = 0$ on and inside the given shape. In Figure 19.1, the boundary is represented as a dark heavy line, and we show the larger shape-offsets obtained using the Fast Marching Method.

Using the Fast Marching Method on triangulated domains, we can

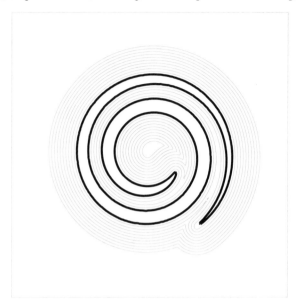

Fig. 19.1. Shape-offsetting of spiral.

extend these ideas to shape-offsetting on surfaces. In [138], the above shape-offsetting was repeated. Figure 19.2 shows shape-offsetting on tori. The original curve is shown in black, while the offsets are shown in white around the curve.

19.2 Voronoi diagrams

Given a collection of seed points, the Voronoi diagram subdivides a domain into different regions. Each region contains all points which are closer to one seed point than any other. There are various ways to construct these diagrams.

We can use the Fast Marching Method to construct Voronoi diagrams on a flat region in a straightforward manner. In order to construct Voronoi diagrams on manifolds, we use the triangulated version. Given the seed points, we first compute the distance from each of the initial given source points simultaneously using a single heap structure and allow one vertex overlap between distance maps from different sources. We then traverse the triangles, and, for each triangle, linearly interpolate the intersection curve between the two different distance maps. For

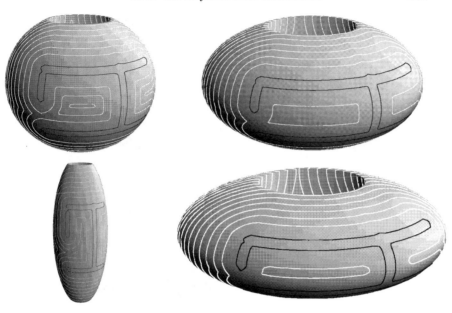

Fig. 19.2. Offsets on four tori.

details of this approach and additional examples, see [138]. In Figure 19.3, we show several Voronoi diagrams on the surfaces of different tori.

19.3 Curve flows with constraints

Next, we study the motion of a constrained interface. This motion is useful in the construction of segments of curves which satisfy various equilibria properties, subject to some other fixed conditions, and has applications in the design of splines, surfaces patches, etc. The approach is straightforward (see [157]). In a small neighborhood of those points where the constraint is desired, the level set function is held fixed (that is, the speed function F is set to zero). This constrains the motion. The use of narrow banding and some re-initialization helps ensure that neighboring level sets do not come together. Figure 19.4 shows two examples of this technique. Starting with an oddly shaped region, pick five points on the initial front, and do not allow these points to move. The points are located at the ends of the white segments. Allowing the front to evolve under curvature flow $F = -\kappa$ produces the polygonal final steady state, with the constraint points at the vertices. Because the curvature is zero along each segment, and the speed function evaluates

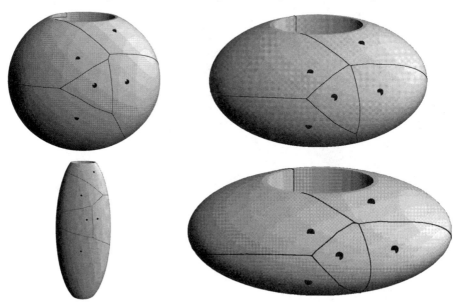

Fig. 19.3. Voronoi diagrams on four tori.

to zero; at all unconstrained points, no additional motion can occur (see Figure 19.4a). On the right (Figure 19.4b), we start with this final polygonal shape and pick five other constraint points. Again, we flow under curvature to produce a polygonal shape, with these new constraint points at the vertices.

19.4 Minimal surfaces and surfaces of prescribed curvature

Consider a closed curve Γ in R^3 with the goal of constructing a membrane with boundary Γ and minimal surface area. In some cases, this can be achieved as follows. Consider some initial surface $S(t = 0)$ whose boundary is Γ. Let $S(t)$ be the family of surfaces (parameterized by t) obtained by allowing the initial surface $S(t = 0)$ to evolve under mean curvature, with the boundary of $S(t)$ always given by Γ. Defining the surface S by $S = \lim_{t \to \infty} S(t)$, one expects that the surface S will be a minimal surface for the boundary Γ. There are several computational approaches for building minimal surfaces based on this approach, including Brakke's Surface Evolver program [36].

Chopp [59] developed a level set approach to this problem by embedding the motion of the surface as the zero level set of a higher dimen-

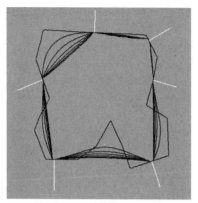
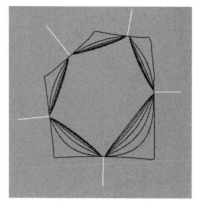

Evolution of initial curve   Evolution of constrained initial polygon
to polygon constrained at vertices   to different polygon
(a)   (b)

Fig. 19.4. Constrained flows.

sional function. Thus, given an initial surface $S(0)$ passing through Γ, construct a family of neighboring surfaces by viewing $S(0)$ as the zero level set of some function ϕ over all of R^3. Using the level set equation (Eqn. 1.5), evolve ϕ according to the speed law $F(\kappa) = -\kappa$. Then a possible minimal surface S will be given by

$$S = \lim_{t \to \infty} \{x | \phi(x,t) = 0\}. \tag{19.2}$$

The difficult challenge with this approach is to guarantee that the evolving zero level set is always attached to the boundary Γ. This is accomplished by creating a set of boundary conditions on those grid points closest to the wire frame that link together the neighboring values of ϕ in order to force the level set $\phi = 0$ through Γ. The underlying idea is most easily explained through a one-dimensional example. Here, we follow the discussion by Chopp [59].

19.4.1 Equations of motion/algorithm

Consider the warmup problem of finding the shortest distance between two points A and B in the plane. The goal is to give conditions on the level set function $\phi(x,t) = 0$ defined in R^2 so that $\phi(A,t) = \phi(B,t) = 0$ for all time.

In Figure 19.5, an initial curve is shown which corresponds to the level set $\phi(x,t) = 0$, together with boundary points A and B. Sup-

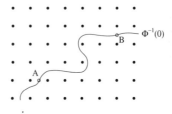

Fig. 19.5. Grid points around the boundary.

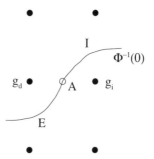

Fig. 19.6. Grid points around the boundary.

pose that the point A is located in between two grid points g_i and g_d (Figure 19.6). In order to force $\phi(A, t) = 0$ for all time, we can require that $\phi(g_i, t) = -\phi(g_d, t)$. Chopp's scheme determines the normal direction, and labels points in that direction as "independent"; points on the other side are hence dependent and controlled by the values for the independent points. More precisely, label the subscripts d and i for dependent and independent, and set the dependent point in terms of the independent point. This binds the dependent points to the independent points in a way that forces the zero level set through the points A and B. In general, the boundary conditions will be represented as a vector equation of the form

$$v_{\text{dep}} = A v_{\text{ind}}, \qquad (19.3)$$

where

$$v_{\text{dep}} = \left\{ \begin{array}{c} \phi(g_{d,1}) \\ \phi(g_{d,2}) \\ \vdots \\ \phi(g_{d,m}) \end{array} \right\} \qquad v_{\text{ind}} = \left\{ \begin{array}{c} \phi(g_{i,1}) \\ \phi(g_{i,2}) \\ \vdots \\ \phi(g_{i,n}) \end{array} \right\}, \qquad (19.4)$$

and A is an $m \times n$ matrix. The matrix A is determined from the chosen mesh and wire frame, and both the classification of dependent and independent points and the matrix A need be computed only once at the beginning of the calculation. This links the set of all dependent points in terms of the set of all independent points in such a way that the level set $\phi = 0$ is forced to pass through the wire frame. Complete details of the automatic technique for generating this list of boundary conditions may be found in [59].

One additional issue comes into play in the evolution of the level set function ϕ toward a minimal surface. By the above set of boundary conditions, only the zero level set $\phi = 0$ is constrained. Thus the other level surfaces are free to move at will, which means that on one side of the level set $\phi = 0$ the surfaces will crowd together, while on the other side they will pull away from the zero level set. This causes numerical difficulties in the evaluation of derivatives for the curvature over such a steep gradient. A re-initialization procedure is used to remedy this situation;[1] after a given number of time steps, the level set $\phi = 0$ is computed, and the function ϕ is re-initialized by directly computing the signed distance function. This uniformly redistributes the level sets so that the calculation can proceed.

19.4.2 Results

We consider a test example of the minimal surface spanning two rings, which has an exact solution given by the catenoid

$$r(x) = a\cosh(x/a), \qquad (19.5)$$

where $r(x)$ is the radius of the catenoid at a point x along the x axis and a is the radius of the catenoid at the center point $x = 0$. Suppose that the boundary consists of two rings of radius R located at $\pm b$ on the x axis. Then the parameter a is determined from the expression

$$R = a\cosh(b/a). \qquad (19.6)$$

If there is no real value of a that solves this expression, then a catenoid solution between the rings does not exist. For a given R, if the rings are closer than some minimal distance $2b_{\max}$ apart, then there are two distinct catenoid solutions, one of which is stable and the other is not. For rings exactly $b_{\max}$ apart, there is only one solution. For rings more than $b_{\max}$ apart, there is no catenoid solution.

---

[1] In fact, this was the first such re-initialization procedure developed for Level Set Methods.

Fig. 19.7. Minimal surface: Catenoid

In Figure 19.7, taken from [59], the minimal surface spanning two rings each of radius 0.5 and at positions $x = \pm 0.277259$ is computed. A cylinder spanning the two rings is taken as the initial level set $\phi = 0$. A $27 \times 47 \times 47$ mesh with space step 0.025 is used. The final shape is shown in Figure 19.7.

Next, in Figure 19.8 (again taken from [59]), this same problem is computed, but the rings are placed far enough apart so that a catenoid solution cannot exist. Starting with a cylinder as the initial surface, the evolution of this surface is computed as it collapses under mean curvature while remaining attached to the two wire frames. As the surface evolves, the middle pinches off, and the surface splits into two surfaces, each of which quickly collapses into a disk. The final shape of a disk spanning each ring is indeed a minimal surface for this problem. This example illustrates a virtue of a level set approach. No special cutting or *ad hoc* decisions are employed to decide when to break the surface. By viewing the zero level set as but one member of a family of flowing surfaces, a smooth transition occurs.

19.5 Extensions to surfaces of prescribed curvature

This technique can be extended to produce surfaces of constant but non-zero mean curvature (see Chopp and Sethian [61]). To do so requires further inspection of the suggestive example of a front propagating with speed $F(\kappa) = 1 - \epsilon\kappa$. Suppose that $\epsilon = 1$, and consider the evolution of the partial differential equation

$$\phi_t + (1 - \kappa)|\nabla\phi| = 0, \qquad (19.7)$$

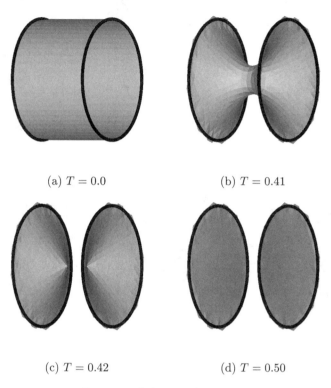

(a) $T = 0.0$ (b) $T = 0.41$

(c) $T = 0.42$ (d) $T = 0.50$

Fig. 19.8. Splitting of catenoid.

where again, the mean curvature is given by Eqn. (6.36), and choose initial data given by

$$\phi(x, y, z, t = 0) = (x^2 + y^2 + z^2)^{1/2} - 1. \qquad (19.8)$$

The zero level set is initially the sphere of radius one, which remains fixed under the motion $F(\kappa) = 1 - \kappa$. All level surfaces inside the unit sphere have mean curvature greater than one, and hence propagate inward, while all level surfaces outside the unit sphere have mean curvature less than one, and hence propagate outward. Thus, the level sets on either side of the zero level set unit sphere pull apart. If one were to apply the level set algorithm in free space, the gradient $|\nabla\phi|$ would flatten out to zero across the unit sphere surface, causing numerical difficulties. However, the re-initialization process described earlier periodically rescales the labeling of the level sets, and thus $|\nabla\phi|$ is renormalized, producing

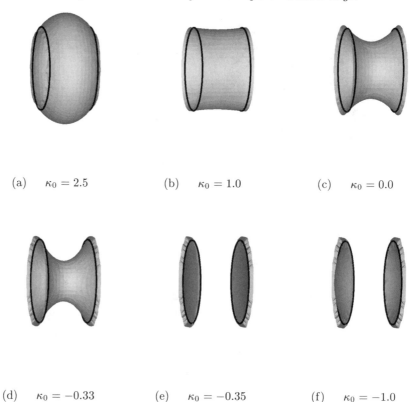

(a) $\kappa_0 = 2.5$ (b) $\kappa_0 = 1.0$ (c) $\kappa_0 = 0.0$

(d) $\kappa_0 = -0.33$ (e) $\kappa_0 = -0.35$ (f) $\kappa_0 = -1.0$

Fig. 19.9. Surfaces of constant mean curvature.

a final surface of constant mean curvature $\kappa = 1$. Thus, in order to construct a surface of constant curvature κ_0, start with any initial surface passing through the initial wire frame and allow it to propagate with speed

$$F(\kappa) = \kappa_0 - \kappa. \qquad (19.9)$$

Here, as before, the "constant advection term" κ_0 is taken as the hyperbolic component F_A and treated using the entropy-satisfying upwind difference solver, while the parabolic term κ is taken as F_B and is approximated using central differences.

Using the two ring "catenoid" problem as a guide (taken from [61]), in Figure 19.9 this technique is used to compute the surface of constant curvature spanning the two rings. In each case, the initial shape is the cylinder spanned by the rings. The final computed shapes are shown for a variety of different mean curvatures. In Figure 19.9(a) a surface of mean curvature $\kappa = 2.50$ is given: the rings are located a distance 0.61 apart and have diameter 1.0. The resulting surface bulges out to fit against the two rings. In Figure 19.9(b) a surface of mean curvature $\kappa = 1.0$ is found, which corresponds to the initial surface. The slight bowing is due to the relatively coarse $40 \times 40 \times 40$ mesh. In Figure 19.9(c) the catenoid surface of mean curvature $\kappa = 0.0$ is given. We isolated the value of $\kappa = -0.33$ as a value close to the breaking point (Figure 19.9(d)). In Figure 19.9(e), a mean curvature value of $\kappa = -0.35$ is prescribed, causing the initial bounding cylinder to collapse onto the two rings and bulge out slightly. Finally, in Figure 19.9(f), bowing out disks corresponding to surfaces of mean curvature $\kappa = -1.00$ are shown. These techniques can be extended to the construction of surfaces of nonconstant curvature. For details, see [59, 61, 60, 133].

19.6 Boolean operations on shapes

In CAD/CAM designs, as well as many other aspects of computational geometry, one often works with a shape descriptor for geometric objects which is based on representations of the boundaries. For example, a polygon in a plane is efficiently represented by a set of ordered vertices; the connectedness of these vertices then produces edges which describe the shape. In three dimensions, multi-sided irregular objects are described in the same hierarchical manner: ordered points yield edges, which in turn yield facets which then connect to describe the given shape. In solid modeling, such operations may need to be performed many times on complex structures.

This is often the most compact way to describe objects and is one of the most common approaches used in a wide collection of software and commercial shape descriptors. Nonetheless, there is a drawback to this approach, namely, that it can be challenging to perform operations on these shapes.

For example, given two closed curves, each described by set of ordered nodes, how does one easily form the union of these curves? Obviously, the goal is to somehow remove those grid points which do not lie on the actual exterior boundary. This can require some effort in two dimensions; in three dimensions, considerable care must be taken to devise an algorithm that deals with all the cases of facets intersecting other facets. It is not easy. Similarly, constructing algorithms to build intersections and subtractions (one shape "minus" another) can be complex.

In the case of performing these Boolean operations on a relatively small number of shapes, this is a workable technique; when large numbers of objects are considered, a different approach is sometimes preferable.

A more precise definition of the problem is as follows. Given two regions Ω_A and Ω_B, the standard operations include forming $\Omega_A \cup \Omega_B$, $\Omega_A \cap \Omega_B$, and $\Omega_A - \Omega_B$. In [236], a technique was introduced for performing Boolean shape operations on implicit surfaces. We now discuss that approach.

19.6.1 Boolean operations on implicit functions

Suppose we form the signed distance functions ϕ_A and ϕ_B for the two regions. Then

- $\Omega_A \cup \Omega_B = \min(\phi_A, \phi_B)$
- $\Omega_A \cap \Omega_B = \max(\phi_A, \phi_B)$
- $\Omega_A - \Omega_B = \max(\phi_A, -\phi_B)$

Our plan is to develop fast algorithms to build these signed distance functions and execute these operations.

We first point out a straightforward way to execute these tasks. Suppose we are given R regions, labeled $k = 1, \ldots, R$, and each described by R_k polygons. We can execute Boolean operations on these shapes by first laying down a grid and then computing the signed distance function ϕ_k on this grid for each region. We can then perform any combination of Boolean operations, for example,

$$[(\phi_1 \cup \phi_3) \cap (\phi_2 \cup \phi_5)] - [(\phi_5 - \phi_2) \cup \phi_1], \qquad (19.10)$$

by parsing from the inside to the outside. Each triple [implicit function, Boolean operation, implicit function] is replaced by another level set function using the min and max operations given above.

We note the following.

- The level of accuracy is determined by the size of the grid. When these Boolean operations are combined with a level set evolution algorithm, the objects already exist on an underlying grid.

- The signed distance functions ϕ_k are reuseable. That is, the expense is in the calculation of the signed-distance functions for each region. Once these signed distance functions are stored, they can be quickly combined under Boolean operations; additionally, new regions then can be added to the operations by calculating their ϕ values and including them in the operations.

- The signed distance functions are easily translated and rescaled. Thus, a library of ϕ values can be quickly translated or rescaled. Thus, the individual objects can be moved around the domain with ease. Rotation is also straightforward; however, some interpolation is required.

- Three-dimensional application is straightforward.

There are two drawbacks to the above algorithm. First, the use of a grid to perform the operations gives an intrinsic resolution scale to the problem. Thus, if the polygon has, for example, long thin spikes, a fine grid may be required to resolve the large aspect ratio. Second, this algorithm is expensive. As a rough operation count, imagine that we have P polygons, each with S sides, and the calculation is performed on an $N \times N$ grid. Calculating each signed distance function requires finding the distance from N^2 points to S sides; thus we require $O(N^2 S P)$ operations to find all the values for ϕ_k. Then, if we perform a simple union, for example, this requires P comparisons of N^2 points, producing a total operation count of $O(N^2 P S + P N^2)$. There are faster ways to do this.

19.6.2 A faster algorithm for Boolean shape operations

A faster algorithm results from exploiting the work on Fast Marching Methods. Construct ϕ for each polygon as follows. For each pair of vertices, construct the line segment connecting the two and locate all grid points that lie on the boundaries of cells that are intersected by the line segment; this is done by following the line through the grid and tagging the mesh points. By respecting the orientation of the line segment, the signed distance function can easily be assigned at the nearby mesh points. This signed distance function is calculated, and the grid point is tagged as having a value for ϕ for this region. The location of this grid point is added to a list of all tagged grid points. One can proceed through all segments, and then through all remaining polygons.

The above (max/min) functions can be executed to produce the Boolean operations, with one extra trick. We perform the operation only at those tagged grid points that are on the list, taking as contributions only those values that have been registered. Values that have not been registered do not contribute to the Boolean operations.

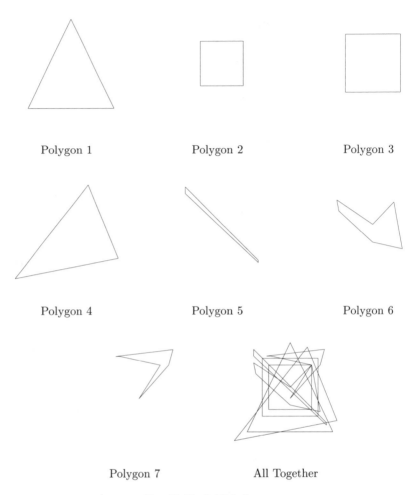

Fig. 19.10. Initial shapes.

A rough operation count can be made on this method. For each side, we tag N points; thus, constructing the signed distance function requires $O(SN)$ operations. Assuming that a total of PSN points are tagged, then the total operation count to compute the union over all P polygons is $O(PSN + PN)$, which is a savings of N^3. All that is required to compute the boundary is a contour plotter that looks for the zero level set in cells where all values exist. The same algorithm extends in a straightforward manner to three dimensions.

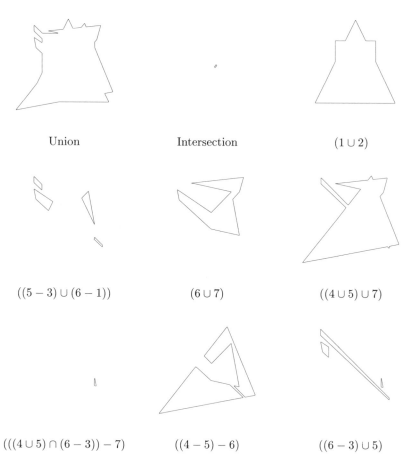

Fig. 19.11. Boolean operations.

19.7 Results: Extracting and combining two-dimensional shapes

Figure 19.10 shows a collection of shapes. Each shape is implicitly defined by means of a signed distance function calculated in a narrow band around the boundary using the Fast Marching Method.

Figure 19.11 shows the results of applying our algorithm to execute a variety of Boolean operations on these shapes. First, ϕ values are constructed for each shape; performing the various Boolean operations is then an extremely fast operation.

19.8 Shape smoothing

Suppose one is given a rough surface. This roughness may be due either to noise in the description of the surface or to oscillations in the shape that one would like to remove. For example, in images reconstructed from range data, oscillations on the surface may be undesirable; similar problems arise with ultrasound data. Another example might be a surface which is produced by experimental apparatus. We would like to extract the essential object and ignore the small-scale oscillations. Our goal is to somehow remove the noise and oscillations on this surface without altering, in a substantial way, the underlying shape.

Before discussing this problem further, we point out, as we did in the section on noise reduction, that there must be some clear decision made about what one is trying to remove before one can just plow ahead. For example, given a surface, one straightforward approach is to run mean curvature flow on an implicitly defined version of the object. This evolves all the level sets under the speed function $F = -\kappa$. As we have seen, this will remove noise, but it will also have the effect of shrinking the entire region.

Suppose we would like to smooth the shape without significantly affecting the enclosed volume. One approach might be to use the volume-preserving flows presented earlier; however, these require the expense of computing the average surface curvature.

Instead, we now describe an algorithm due to Malladi and Sethian [160] which exploits the Min/Max flow described in the previous chapter on image processing. Recall the Min/Max flow (Eqn. 16.6), namely

$$\bar{F}_{min/max} = \left\{ \begin{array}{ll} \max(\kappa, 0) & \text{if Ave}_{\phi(x,y)}^{R=kh} < T_{\text{threshold}} \\ \min(\kappa, 0) & \text{otherwise} \end{array} \right\}. \qquad (19.11)$$

The threshold determines the size of the smoothing level. It is geometrically based on physical space. It removes oscillations on the order of the threshold and then stops. We show two examples of this shape smoothing. In Figure 19.12, we see that mean curvature flow shrinks the basic underlying shape as the oscillations are removed, while the Min/Max flow tries to maintain the underlying shape. Figure 19.13 shows Min/Max flow applied to a noisy ultrasound image.

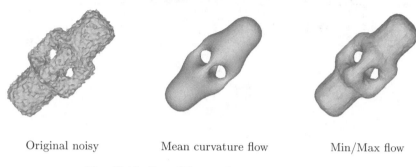

Original noisy   Mean curvature flow   Min/Max flow

Fig. 19.12. Two different shape smoothing flows.

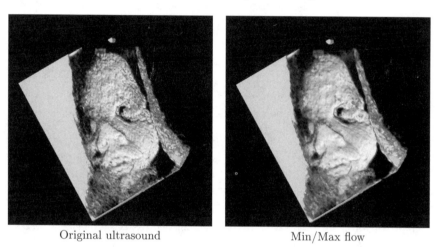

Original ultrasound   Min/Max flow

Fig. 19.13. Min/Max shape smoothing flow applied to ultrasound image.

20

Optimality and First Arrivals

Outline: *The Eikonal equation, which gives the first arrival or viscosity solution, is at the core of several applications involving optimization. In this chapter, we explore some examples. The main tool in all our applications will be the Fast Marching Method. We discuss problems in path planning, computation of first arrivals in seismic traveltimes, constructing shortest geodesic paths, control, computing visibilities, and line-of-sight evaluations.*

One aspect of viscosity solutions is that they extract, among all possible solutions, the one that corresponds to the first arrival of information from the initial disturbance. This is in keeping with the view that the information can be traced back to the initial data via the shortest path. This leads to a set of applications of our algorithm methodology designed to produce optimal solutions, based on minimum distances or shortest arrival times.

20.1 Optimal path planning

As a first application, we use the Fast Marching Method to solve problems in path planning. Here, we follow the approach laid out by Kimmel and Sethian [135]. Given initial and final positions of an object and a domain which may contain obstacles, the goal is to find the shortest possible path from the initial state to the final state. The domain may have a weighted metric on it, so that it might cost more to traverse some areas than others; depending on the application under study, the object may also be free to rotate as it moves. As an example, consider a L-shaped object moving in a two-dimensional area, with several impenetrable and oddly shaped obstacles in the region. If the object is free to

284

rotate as it moves, then this is a problem with three degrees of freedom. One says that the configuration space is three-dimensional. If there is a movable arm attached to the robot, then the problem has a four dimensional configuration. A robot moving in three-dimensional space exists in a six-dimensional configuration space;[1] including arm motion adds significantly greater complexity. Because our techniques are based on laying down a grid in configuration space, without modification they are restricted to problems with relatively low degrees of freedom.

20.1.1 Statement of problem

Given a cost function $F(x_1, x_2, .., x_n)$ and a starting point A in R^n, the goal is to find the path $\gamma(\tau) : [0, \infty) \to R^n$ from A to any point B in R^n which minimizes the integral

$$\int_{A=\gamma(0)}^{B=\gamma(L)} F(\gamma(\tau))d\tau, \tag{20.1}$$

where τ is the arc length parameterization of γ, namely, $|\gamma_\tau| = 1$, and L is the total length of γ. More specifically, in two dimensions, suppose we are given the cost function $F(x, y)$ and starting point A. Let $T(x, y)$ be the minimal cost required to travel from A to the point (x, y), that is,

$$T(x, y) = \min_\gamma \int_A^{(x,y)} F(\gamma(\tau))d\tau, \tag{20.2}$$

with $|\gamma_\tau| = 1$. The level set $T(x, y) = C$ is the set of all points in R^2 that can be reached with minimal cost C and the minimal cost paths are orthogonal to the level curves. Thus, we have

$$|\nabla T| = F(x, y). \tag{20.3}$$

Our goal is to apply the Fast Marching Method to solve this Eikonal equation to produce $T(x, y)$ in all of R^2. Then, given a point B in R^2, explicit construction of the shortest path comes through back propagation from B to A via the solution of the ordinary differential equation

$$X_t = -\nabla T, \qquad \text{given} \qquad X(0) = B, \tag{20.4}$$

until we reach the starting point A.

More precisely, given the arrival times T, we can construct the optimal path by tracing backward from the final configuration to the initial

---

[1] Three coordinates for position and three for orientation of the body.

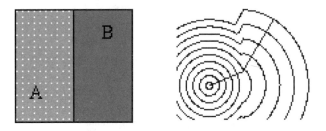

Fig. 20.1. Two-dimensional navigation on variable domain.

one by solving the ordinary differential equation given in Eqn. 20.4. We do so by using second order Heun's method to integrate the ordinary differential equation. Starting from an initial point P_o, we take one step using Euler's method to compute a temporary value. We then evaluate ∇T at this point by bilinear interpolation from the values of the gradient at each corner of the cell which contains the particular. These gradient values at cell corners are found by straightforward upwind differences using neighboring values. Once the gradient is evaluated at the temporary point, Heun's method is used to construct an average gradient. This is used to advance the position of the point, and the process is repeated.

20.1.2 Results

20.1.2.1 Two degrees of freedom

We begin with a straightforward application of the technique to two-dimensional path planning with constraints. In Figure 20.1 we show the optimal path from point A to point B in a dual-valued domain; the speed F on the left is half the value of the speed on the right. On the left, we show the initial and end points, together with the two domains; on the right, we show the equal time T solutions obtained using the Fast Marching Method together with the optimal path obtained by solving the ordinary partial differential equation. The results, as expected, show the effects of Snell's law in the shortest path between two differing media.

In Figure 20.2 we add constraints to the motion in the form of impenetrable rectangles. We set the speed function F to zero inside these regions, and the corresponding alteration in the optimal path is shown.

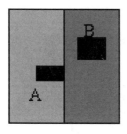 

Fig. 20.2. Two-dimensional navigation with constraints on variable domain.

20.1.2.2 Three degrees of freedom

We next turn to two-dimensional problems with three degrees of freedom. Consider now the problem of navigation with constraints and rotation. Instead of a robot as a single point, we imagine a two-dimensional rectangle with a given width and length; thus the initial position A of the robot in configuration space is specified by the position of the center of the rectangle, plus an angle θ between 0 and 2π. The final configuration B is similarly specified, and the goal is to construct the optimal path from A to B.

In the absence of obstacles, a completely straightforward application of the Fast Marching Method is possible. We discretize the configuration space into a three-dimensional grid, that is, we grid both R^2 and θ between 0 and 2π, employing periodic boundary conditions in θ. Thus, we solve the Eikonal equation

$$\left[u_x^2 + u_y^2 + u_\theta^2\right]^{1/2} = 1. \tag{20.5}$$

In the presence of obstacles, we take the following approach (see [144]). Rather than maneuver an oddly shaped robot, we instead consider the robot as a point and, for every discretized angle θ_i, alter the shapes of the obstacles corresponding to that angle. For example, imagine that one of the obstacles is a square of unit width and height in the domain. Suppose the rectangular robot has length .5, and consider the discretization of configuration space at the angle $\theta_i = 0$. Then, in this direction the center of the robot can come only within 0.25 of the square in the positive or negative x direction; hence we may consider, for this angle, the equivalent problem of a point robot avoiding an obstacle which has height 1 and length 1.5.

To do this for all obstacles at all angles requires morphological shape

Fig. 20.3. Two-Dimensional navigation around obstacles with rotation.

operations consisting of dilations and translations. In some cases, this can be done analytically; in other cases, the construction of obstacles in configuration space can itself be performed using the Fast Marching Method. For example, for a robot shaped as a ball of radius r, we can use the method to compute the distance T^0 from the obstacles, and we thereby obtain the extended obstacle space $T^0 < r$ and the dilated free space $T^0 > r$ in which the robot is considered a free point. In Figure 20.3 we show several examples of a two-dimensional robot with rotational angle navigating in two-dimensional space around obstacles.

20.1.2.3 Four Degrees of Freedom

Finally, in Figure 20.4, we consider a motion problem which contains four degrees of freedom. We consider a stick fixed at the lower point, with three hinged segments. The goal is to move the stick from one position (shown in light gray) to the other position. Figure 20.4 shows the intermediate stages, left to right and top to bottom, as the stick is moved from the initial position to the final position. We refer the reader to the original reference [135] for further examples and timings.

20.2 Constructing shortest paths on weighted domains

20.2.1 Shortest paths

The above path planning problem solves the Eikonal equation

$$|\nabla T| = \frac{1}{F(x, y, \theta)},$$

in which the speed function F is binary. it is one in the reachable regions and 0 inside the obstacles.[2]

There is no reason why we must limit ourselves to binary speed values. Thus, for continuously varying F, we may use the same algorithm to compute the shortest path on a domain with a weighted metric. Once again, we expand a front using the Fast Marching Method, compute the arrival times, and then trace backward along the gradient of the arrival time function to produce the shortest path.

As a simple example, taken from [134], Figure 20.5 shows a domain of 100×100 grid points in which the speed in the light gray region is $F(x, y) = 1$ for $i < 60$, and in the dark gray region the speed is $F(x, y) = 5$ for $i \geq 60$. Thus, there is a speed jump of a factor of five between the left and right sides. A disturbance is started in the slower region at the point $(30, 10)$, and the equal time contours are computed, together with the paths from a set of points reaching back to the initial disturbance.

---

[2] Recall how the Fast Marching Method works; a speed of $F = 0$ means that it takes an infinite amount of time to propagate through an obstacle, and hence the $FarAway$ value is equivalent to never reaching the point.

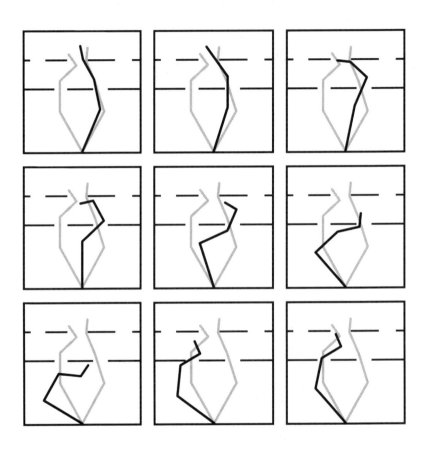

Fig. 20.4. Optimal motion involving four degrees of freedom.

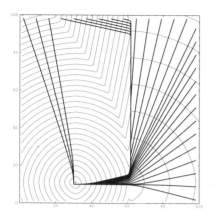 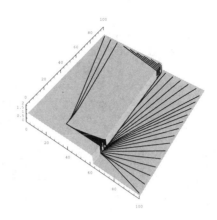

(a) Shortest paths for on a weighted domain   (b) A perspective view

Fig. 20.5. Computation of shortest paths on weighted domains; a two-valued media.

Figure 20.5(a) shows these paths drawn on the equal time contours. Note that the two sets of curves are perpendicular to each other as expected. In Figure 20.5(b), we plot the path lines on a perspective view where the height corresponds to the speed. As can be seen, some of the paths in the slow region reach back to the initial disturbance by first moving over to the fast region, tracking back along the boundary between the fast and slow, and then reaching back to the initial disturbance.

20.2.2 Shortest paths with bounces

We may extend this algorithm to compute the shortest path between two points that includes a given number of bounces from solid walls or internal structures. Suppose we are given two points A and B in a closed domain with weight function F prescribed everywhere throughout the domain. Figure 20.6 shows two shortest paths for a particular weight function F; the first is the true shortest path (path 1), and the second is the shortest path (path 2) that contains a bounce off of a wall. We can compute path 1 by using the above Fast Marching Method. To compute path 2, we first compute the arrival time T for the entire domain. We then let the arrival time values along the boundary serve as boundary conditions for a *new* application of the Fast Marching Method, providing the shortest single bounce path from A to B. Repeated applications of this technique finds multi-bounce shortest paths. This algorithm is useful for computing reflected bounces and arrivals off of internal and external structures. For details about this technique and more sophisticated algorithms to compute multivalued solutions of the Eikonal equation, see [240].

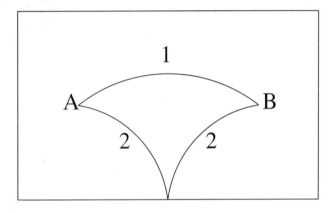

Fig. 20.6. Finding shortest paths with bounces.

20.3 Constructing shortest paths on manifolds

Suppose we now try to compute the shortest paths to an initial point on manifolds. Here, the issue is that the domain itself is not flat, and hence we must solve the problem of a front propagating with speed $F = 1$ on the curved surface itself. This is a problem of considerable interest. It is connected, for example, to aspects of computer graphics, antenna placement and wave propagation on bodies, and surface grid generation. Two approaches to this problem are possible in our front propagation framework.

20.3.1 A flat domain approach

Consider a surface $z(x_1, x_2, .., x_M)$. The objective is to find the shortest path, known as a minimal geodesic, between two points $(A, z(A))$ and

$(B, z(B))$ on that surface. Here, the discussion and results of Kimmel and Sethian [134] are presented almost without change. Imagine the unit speed propagation of the initial surface curve $\Gamma(t)$ along the surface, where $\Gamma(t = 0)$ corresponds to the single point $(A, z(A))$. The location of the curve at time t will correspond to the set of all points of distance t from the point $(A, z(A))$, and again, the shortest path will correspond to back propagation along the surface vector field orthogonal to the surface curves Γ.

Our goal is to reframe the motion of the front on the manifold in terms of its motion in the domain space $(x_1, x_2, \ldots, x_M)$. For simplicity, we discuss the problem for a surface which maps R^2 into R. Since the surface $z(x, y)$ is a graph, there exists some speed function $\tilde{F}$ which provides the corresponding motion of the surface curves projected onto the $x - y$ plane. The goal is to determine this speed function $\tilde{F}$.

We first note that the evolution on the surface itself before the projection is given by

$$\Gamma_t = N \times T, \tag{20.6}$$

where, using the notation $(p, q) = \nabla z$, the surface normal is $N = (-p, -q, 1)/\sqrt{1 + p^2 + q^2}$, and T is the tangent to the current equal geodesic distance contour Γ. This evolution can be used to compute the geodesic distance map from a given point or a set of points on the surface $z(x, y)$. Thus

$$\frac{1}{\tilde{F}(\nabla z, n)} = (\Pi \circ N \times T, n), \tag{20.7}$$

where $\Pi \circ (x, y, z) = (x, y)$ is the projection operation onto the $x - y$ coordinate plane.

It can be shown (see [134]) that the tangent T is given by

$$T = \frac{(-u_y, u_x, qu_x - pu_y)}{\sqrt{u_x^2 + u_y^2 + (qu_x - pu_y)^2}}.$$

Since $n = \nabla u / |\nabla u|$, one can solve for $|\nabla u|$ to produce a static Hamilton–Jacobi equation of the surface distance map u, namely,

$$|\nabla u|^2 = \tilde{F}(\nabla z, n)^2. \tag{20.8}$$

The result is as the solution of

$$\frac{u_x^2(1 + q^2) + u_y^2(1 + p^2) - 2pqu_xu_y}{1 + p^2 + q^2} = 1. \tag{20.9}$$

with boundary conditions $u = 0$ at A. This results in the following Hamiltonian:

$$H(u_x, u_y) = (1+q^2)u_x^2 + (1+p^2)u_y^2 - 2pqu_xu_y - (1+p^2+q^2). \quad (20.10)$$

Again, the objective is to find the surface u that solves the static Hamilton–Jacobi equation given by $H(u_x, u_y) = 0$, with the boundary conditions $u = 0$ at A.

Finally, given the solution u to the static Hamilton–Jacobi equation, we need to construct the actual geodesics by back propagating from the point B to the starting point A by solving the ordinary differential equation

$$X_t = -\mathbf{\Pi} \circ (N \times T) \qquad \text{given} \qquad X(0) = B. \qquad (20.11)$$

Since N is given as a function of ∇z, and T is obtained by back projecting the level set of u onto z, substitution into the above expression becomes

$$X_t = -\frac{\left(u_x(1+q^2) - pqu_y, \, u_y(1+p^2) - pqu_x\right)}{\sqrt{(1+p^2+q^2)(u_x^2 + u_y^2 + (qu_x - pu_y)^2)}}. \qquad (20.12)$$

This is a reasonable approach. In many cases, it can be used together with the Fast Marching Method to produce a consistent viscosity scheme which satisfies the causality update condition. However, one can easily imagine a situation in which this is not easy to do. Imagine a surface or shape such that motion along a diagonal is far preferable to motion along any of the grid lines. This mean that the diagonal point is reached first, and hence it is impossible to correctly compute the arrival time at the diagonal point if it relies on earlier and incorrect values at the neighboring grid points. Thus, we must use an alternative approach.

20.3.2 An $O(N \log N)$ method for constructing shortest geodesic paths on manifolds

The alternative approach, as first presented in [137], is to use the triangulated Fast Marching Method discussed earlier to solve the straightforward Eikonal equation $|\nabla T| = 1$ on a triangulated approximation to the manifold itself. We then backtrace on the surface itself, solving the ordinary differential equation

$$\frac{dX(s)}{ds} = -\nabla T,$$

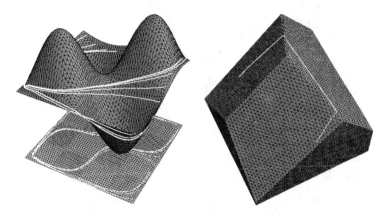

Fig. 20.7. Computing minimal weighted geodesics on triangulated surfaces.

where $X(s)$ traces out the geodesic path. We use a second order Heun's integration method on the triangulated surface with a switch to a first order scheme at sonic points in the gradient. When integrating within a given triangle, its three neighboring triangles support the computation and are used to interpolate T as a second order polynomial whose six coefficients are computed from those given T values at the vertices.

Here, we repeat the examples laid out by Kimmel and Sethian in [137]. Figure 20.7 presents a perspective view of the triangulation and shortest paths for two surfaces. Figure 20.7a shows shortest paths on regular triangulation of the surface given by the function $z(x, y) = 0.45 \sin(2\pi x) \sin(2\pi y)$ on $[0, 1] \times [0, 1]$, for grid size of 50×50. The minimal geodesics are painted on the triangulated surface and projected to the $x - y$ plane. Figure 20.7b gives a polyhedron example, in which a different speed function F was assigned to each side, causing a Snell law effect along the edges. The speed F is 2 at the close side with the start point, 1 at the second top side, and 4 at the side of the destination point.

Next, we show two additional computations to illustrate the performance on manifolds with an underlying non-acute (obtuse) triangulation. Figure 20.8 shows the computation of minimal geodesics on a torus genus one object, in which some shortest paths in fact cut through the middle. The equi-distance curves are shown, as well as the shortest paths. In Figure 20.9, we compute geodesic distances on a synthetic head.

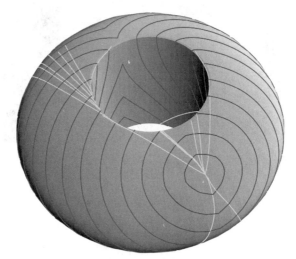

Full view of shortest paths

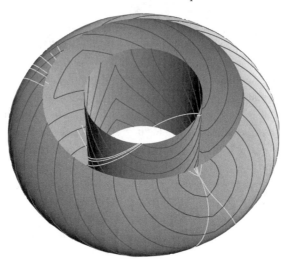

Cutaway view

Fig. 20.8. Shortest paths on a bead (a genus one 2D manifold).

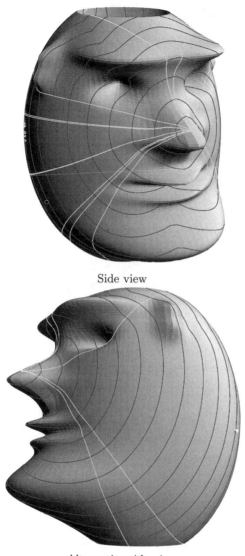

Side view

Alternative side view

Fig. 20.9. Shortest paths on synthetic head.

We note that the computational complexity is unchanged; we have given an $O(N \log N)$ algorithm for constructing geodesic paths on triangulated surfaces. To the best of our knowledge, it is the fastest such algorithm;[3] for details, see [137].

20.4 Seismic traveltimes

Next, we explore applications of Fast Marching Methods to problems involving the imaging of geophysical data sets. In [242], Sethian and Popovici used the Fast Marching Method to rapidly construct first ar- rival times in seismic analysis and then coupled this work to prestack migration. Here, we summarize that work. For further details, see [242].

Three-dimensional (3D) prestack migration of surface seismic data is a tool for imaging the earth's subsurface when complex geological struc- tures and velocity fields are present. The most commonly used imaging techniques applied to 3-D prestack surveys are methods based on the Kirchhoff integral, because of its flexibility in imaging irregularly sam- pled data and its relative computational efficiency. In order to perform this Kirchhoff migration, one approximately solves the wave equation with a boundary integral method. The reflectivity at every point of the earth's interior is computed by summing the recorded data on mul- tidimensional surfaces; the shapes of the summation surfaces and the summation weights are computed from the Green's functions of the sin- gle scattering wave-propagation experiment (see [220, 200]).

20.4.1 Background equations

In some more detail, the essence of 3-D prestack migration is expressed by the following integral equation:

$$\text{Image}(\mathbf{x}) = \int \int_{\mathbf{x_s}} \int_{\mathbf{x_r}} G(\mathbf{x_s}, \mathbf{x}, \omega) G(\mathbf{x}, \mathbf{x_r}, \omega) \text{Data}(\mathbf{x_s}, \mathbf{x_r}, \omega) d\mathbf{x_r} d\mathbf{x_s} d\omega,$$

where $\mathbf{x}$ is the image output location, and $\mathbf{x_s}$ and $\mathbf{x_r}$ are the data source and receiver coordinates, and ω is angular frequency. The Green's func- tions $G(\mathbf{x_s}, \mathbf{x}, \omega)$ and $G(\mathbf{x}, \mathbf{x_r}, \omega)$ parameterize propagation from source to image point and from image point to receiver, respectively. In most

---

[3] It is important to be clear about what is being claimed. We are *not* finding the exact shortest path for a given triangulation of a surface. The solution path using the above algorithm is the shortest path within an error which depends on the size of the triangulation. It is an *approximate* shortest path, constructed in $O(N \log N)$ steps, where N is the number of triangles used to tessellate the surface. As the triangulation is refined, the path converges to the exact shortest path.

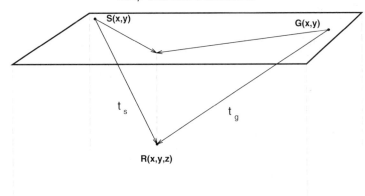

Fig. 20.10. In 3-D seismic surveying, seismic waves are generated by surface sources (shots) S, and the reflected waves are recorded at surface receivers (geophones) G. The Green's function describes the energy of the wavefield back-scattered from the reflector point at all possible source and receiver combinations.

implementations, the calculation is often done instead in the time domain, and can be expressed as the summation

$$\text{Image}(\mathbf{x}) = \sum_{\mathbf{x_s}} \sum_{\mathbf{x_r}} A_s A_r \text{Input}(\mathbf{x_s}, \mathbf{x_r}, t_s + t_r),$$

where Input is a filtered version of the input data, and the Green's functions are parameterized by the amplitude terms A_s and A_r and traveltimes t_s and t_r.

For 3-D prestack Kirchhoff depth migration, the Green's functions are represented by five-dimensional (5D) tables; these tables are functions of the source/receiver surface locations (x, y) and of the reflector position (x, y, z) in the earth's interior. This Green's function parameterization is usually based on the assumption of acoustic propagation. This Kirchhoff prestack migration process consists of two stages. First, traveltime tables are computed and stored. Second, the migrated image is formed by convolving the prestack data with migration operators derived from the traveltime tables. Both phases present challenges from the perspective of the geophysical accuracy and of the computer implementation (see Figure 20.10).

The key element of 3-D prestack Kirchhoff depth migration is the calculation of traveltime tables used to parameterize the asymptotic Green's functions. An efficient traveltime calculation method is required to generate the 5-D traveltime tables needed for 3-D Kirchhoff migration.[4] Also, since depth migration problems are generally applied in areas of complex velocity structure, the traveltime calculation method must be robust. Computing 3-D Green's function tables over a 100×100 kilometer area (about 430 marine blocks), with sources positioned every 200 meters, requires 1 terabyte of traveltime volumes. Thus, speed is an important issue.

Designing efficient and accurate traveltime computation methods has a long history; the past ten years have seen considerable new advancements, particularly those aimed at a finite difference approach (see Vidale ([277])). Prior to this work, traveltimes were typically computed using ray tracing. While these ray tracing methods offer a high degree of accuracy, they also pose interpolation problems in shadow areas and areas where multiple caustics develop. The use of finite difference traveltimes ameliorates these interpolation problems in shadow zones, at the price of foregoing detection of most energetic arrivals in exchange for the first arrival.

A broad spectrum of traveltime computation methods was developed in the early 1990's. Vidale extended his finite difference traveltime to three dimensions [278], while van Trier and Symes [276] introduced a two-dimensional explicit finite-difference method with a vectorizable inner kernel that ran efficiently on vector computer platforms. At its core, the problem of computing first arrival times requires solution of the Eikonal equation, with the goal of accurately and robustly dealing with the formation of cusps and corners, topological changes in the solution, and singularities. Fast Marching Methods, both first and second order versions, provide viable approaches.

20.4.2 Computing Fast Marching Method traveltimes through a salt structure

We begin by showing the results of using the Fast Marching Method to compute three-dimensional travel times through a salt structure.[5] We start with the techniques applied to a 3-D SEG/EAGE salt dome velocity model [12]. The salt dome model was designed to contain major complex features that are characteristic of complicated Gulf of Mexico salt structures. It includes a northwesterly plunging stock, a secondary

---

[4] The Green's function can be reconstructed from traveltime tables that describe traveltimes from all surface points (x, y) to all subsurface locations (x, y, z); thus the tables are five-dimensional.

[5] All seismic calculations were performed using the implementation of the Fast Marching Method developed by 3DGeo Corporation.

reactivation crest southward of the stock, a low-relief eastern flank, a faulted southern flank with a toe thrust, a rounded overhang on the west flank, five sands that are gas charged (at least one contains both a gas/oil contact and an oil/water contact), and a shale sheath that is modeled to be geopressured. The sea floor map exhibits a counter-regional fault scarp, a bathymetric rise associated with the sill crest, and a shelf break at the southeast end of the model. The overall model size is $13.5 \times 3.5 \times 4.2$ kilometer on a 20 meter grid.

The SEG/EAGE Salt Model has a complicated salt-to-sediments interface which creates complex wave propagation problems. We show contour traveltimes superimposed on the velocity model which are representative for the wave propagation patterns encountered while solving the Eikonal equation in the SEG/EAGE Salt Model. Figure 20.11 shows a traveltime slice through the SEG/EAGE Salt Model with a point source at the surface. The grid is a $100 \times 100 \times 100$ mesh, with mesh size equal to 40 meters per cell side. Figure 20.11b shows the formation of headwaves which travel along the salt-sediment interface. Figure 20.12 shows the result of a horizontal traveltime slice through the traveltime cube at a depth of 1380 meters and accurately captures the formation of cusps.

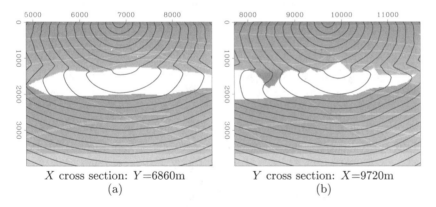

| X cross section: $Y=6860$m | Y cross section: $X=9720$m |
| (a) | (b) |

Fig. 20.11. Traveltime slices through SEG/EAGE Salt Model.

20.4.3 Migration using the Fast Marching Method

Figures 20.13 and 20.14 show slices through the three-dimensional velocity and corresponding structural images obtained from migration on prestack data obtained from a given data set. On the left, Figure 20.13

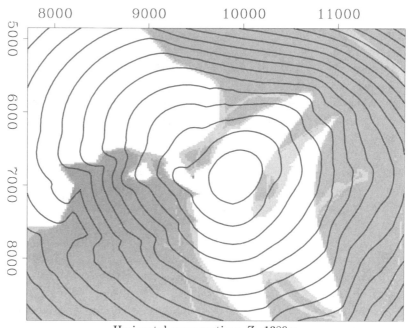

Horizontal cross section: Z=1380m

Fig. 20.12. Horizontal slice through SEG/EAGE Salt Model.

shows a depth slice through a velocity cube at a depth of 1220 meters; on the right, the corresponding migrated image slice is shown. The salt/sediment interface and the semicircular fault cutting through the salt body are imaged with high resolution. Figure 20.14 compares the velocity model on the left with the corresponding migrated line on the right for a different slice. The sediment images are imaged at the correct locations, together with the salt body borders. The areas with lesser quality are under the salt, most probably because of the multiple reflected arrivals at this spot from the water bottom and intra-salt reflections, and also close to the left side of the top of the salt, most probably also the results of the use of first arrivals in the Fast Marching Method.

Next, two-dimensional traveltimes were used to image the Marmousi data, which is a synthetic data set based on a real geologic model from the Cuanza basin in Angola [32]. The geologic model of the basin consists of a deltaic sediment interval deposited upon a saliferous evaporitic series. The sediments are affected by normal growth faults caused by

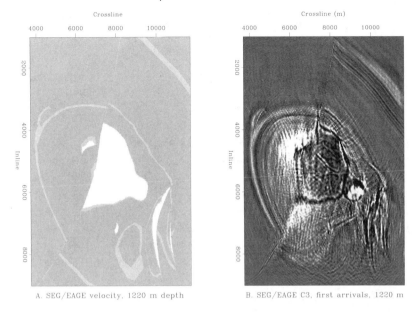

A. SEG/EAGE velocity, 1220 m depth   B. SEG/EAGE C3, first arrivals, 1220 m

Fig. 20.13. Velocity model and and migrated image.

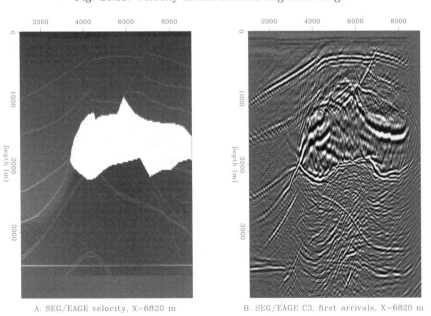

A. SEG/EAGE velocity, X=6820 m   B. SEG/EAGE C3, first arrivals, X=6820 m

Fig. 20.14. Velocity model and migrated image.

the salt creep. Under the salt there is a folded carbonate sedimentation series forming a structural hydrocarbon trap. The challenge is to image the hydrocarbon trap. The complex velocity model, with strong lateral velocity variations, is shown in Figure 20.15 on the left. On the right, the figure shows the migrated images using three-dimensional traveltime tables computed with the Fast Marching Method, operating in a two-dimensional mode. The typical challenges in imaging this dataset are (1) imaging correctly and without artifacts the position of the faults, (2) imaging the V-shaped termination of the layers, which are zones that concentrate rays and produce distorted images, (3) imaging correctly the top of the first anticlinal structure and the bottom of the two salt intrusions, and (4) imaging the sediments in the second, deep anticlinal structures. The images shown correctly image the faults, even in the high velocity layers, and avoid artifacts at the bottom of the V-shaped fault and layer terminations. The fact that the Fast Marching Method produces first arrivals, which may not correspond to the most energetic arrivals, may explain why the second, deeper anticlinal area is missed. For further discussion of this and other features, see [242], as well as [268].

A. Marmousi velocity

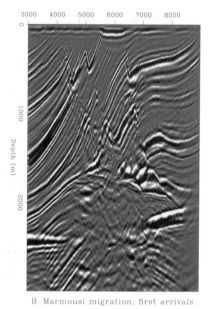

B. Marmousi migration, first arrivals

Fig. 20.15. Velocity model and migrated image, Marmousi data set.

20.5 Aircraft collision avoidance using Level Set Methods

An interesting application of Level Set Methods has been developed by Tomlin [272] to compute solutions to discontinuous Hamilton-Jacobi equations which model the relative motion of two aircraft in collision avoidance maneuvers. Here, we briefly summarize the work in [272]. Consider, for example, the two-aircraft lateral maneuver shown in Figure 20.16. As soon as the aircraft are within a certain distance of each other, each aircraft turns 90° to its right and follows a half circle. Once the half circle is complete, each aircraft returns to its original heading and continues on its straight path. Current Federal Aviation Administration restrictions dictate that aircraft must maintain a 5-mile lateral separation between each other. Also, each aircraft may alter its linear and angular velocities (control inputs) to avoid collision: it is assumed that the control actions of aircraft 2 are *uncertain*, and the goal is to compute the control actions of aircraft 1 such that separation is maintained despite this uncertainty. To do this, Tomlin modeled the system as a dynamic game between two players: the control u, which models the actions of aircraft 1, and the disturbance d, which models the actions of aircraft 2.

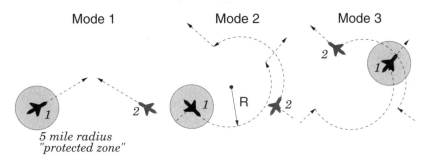

Fig. 20.16. Two aircraft in three modes of operation: in modes 1 and 3 the aircraft follow a straight course, and in mode 2 the aircraft follow a half circle.

The maneuver is modeled as in Figure 20.17; $(x_r, y_r, \psi_r) \in R^2 \times [-\pi, \pi]$ represents the relative position and orientation of aircraft 2 with respect to aircraft 1. The continuous state space is augmented with a timer $z \in R$ to force the transition between the second and third modes, thus we let $x = (x_r, y_r, \psi_r, z)$.

The goal is to determine all initial states x of aircraft 2 which are *unsafe*, meaning that even if aircraft 1 tries to do its best to avoid conflict, aircraft 2 has a trajectory from this state for which it will violate

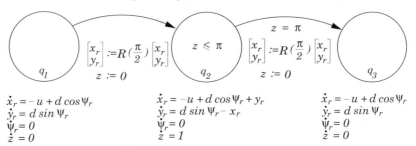

Fig. 20.17. In q_1 the aircraft follow a straight course; in q_2 the aircraft follow a half circle; in q_3 the aircraft return to a straight course.

the separation requirement. To do this, Tomlin back-propagates a 5 mile radius disk from the origin through the dynamics of three modes of the maneuver. This corresponds to evolving the boundary of the disk under the Hamilton-Jacobi equation

$$-\frac{\partial J(x,t)}{\partial t} = \min\{0, \frac{\partial J(x,t)}{\partial x} \cdot f(x, u^*, d^*)\}. \qquad (20.13)$$

Here, $J(x,t)$ is the cost of a trajectory starting at state x at time t (free) and evolving to time 0, the final time, defined so that $J(x,t)$ is negative if the final state lies inside the disk, and is positive if the final state lies outside the disk. Here, $f(x, u^*, d^*)$ is the vector field describing the continuous dynamics in each mode and (u^*, d^*) are the possibly discontinuous optimal control and worst disturbance actions of the two aircraft obtained by solving the dynamic game that are functions of x and $\frac{\partial J(x,t)}{\partial x}$. The "min" in Eqn. 20.13 ensures that once a state becomes unsafe, it stays unsafe.

At any time t, the set $\{x \mid J(x,t) \geq 0\}$ comprises those states from which aircraft 1 can keep the system state outside of the 5 mile disk for at least $|t|$ seconds, and the set $\{x \in X \mid J^*(x,t) < 0\}$ is the set of states from which aircraft 2 can force the system into the 5 mile disk in at most $|t|$ seconds. Tomlin applies the Level Set Method to compute the boundary of these sets $\{x \mid J(x,t) = 0\}$.

The calculation requires switching between modes 1, 2, and 3. Tomlin computes the set $\{x \mid J(x,t) = 0\}$ in each mode (with different continuous dynamics $f(x, u^*, d^*)$), and then "iterates" between modes, determining when aircraft 1 should switch modes so that the unsafe set is minimized. The details of the algorithm are described in [272]; here we present the only the level set algorithm used to calculate $\{x \mid J(x,t) = 0\}$

for a single mode and for a two-dimensional example (state variables are (x_r, y_r) assuming that the heading ψ_r is constant):

Choose a domain of interest in x and discretize the domain with a grid of spacing $\Delta x_1, \Delta x_2$. Let x_{ij} represent the grid point $(i\Delta x_1, j\Delta x_2)$ and let $\tilde{J}(x_{ij}, t)$ represent the numerical approximation of $J(x_{ij}, t)$. Using the boundary condition $J(x, 0) = l(x)$, where $l(x) = x_r^2 + y_r^2 - 25$, compute $\tilde{J}(x_{ij}, 0)$ for each x_{ij}.

Let $t = 0$. While $\tilde{J}(x_{ij}, t) \neq \tilde{J}(x_{ij}, t - \Delta t)$, perform the following steps:

(i) Compute

$$u^*(x_{ij}, \frac{\partial \tilde{J}(x,t)}{\partial x_1}, \frac{\partial \tilde{J}(x,t)}{\partial x_2}) \qquad d^*(x_{ij}, \frac{\partial \tilde{J}(x,t)}{\partial x_1}, \frac{\partial \tilde{J}(x,t)}{\partial x_2})$$
$$(20.14)$$

using the central difference approximations D^{0x_1}, D^{0x_2},

$$\frac{\partial \tilde{J}(x,t)}{\partial x_1} = D^{0x_1}, \qquad \frac{\partial \tilde{J}(x,t)}{\partial x_2} = D^{0x_2}. \qquad (20.15)$$

(ii) Calculate $f(x_{ij}, u^*, d^*)$.

(iii) Depending on the sign of $f(x_{ij}, u^*, d^*)$, use either the forward or backward difference approximations to the partial derivatives $\frac{\partial \tilde{J}(x,t)}{\partial x_1}$ and $\frac{\partial \tilde{J}(x,t)}{\partial x_2}$.

(iv) Compute $\tilde{J}(x_{ij}, t - \Delta t)$:

$$\tilde{J}(x_{ij}, t - \Delta t) = \begin{cases} J(\tilde{x}_{ij}, t) + \Delta t \frac{\partial \tilde{J}(x_{ij},t)}{\partial x} f(x_{ij}, u^*, d^*) \\ \qquad \text{if } \frac{\partial \tilde{J}(x_{ij},t)}{\partial x} f(x_{ij}, u^*, d^*) < 0 \\ \tilde{J}(x_{ij}, t) \text{ otherwise.} \end{cases}$$
$$(20.16)$$

The resulting unsafe set, incorporating the switching between modes, is shown as the shaded region in Figure 20.18. For details and further discussion of differential games and Level Set Methods, see [272].

20.6 Visibility evaluations

Our final application is the evaluation of visibility issues. Here, both the idea of an implicit representation of an interface and Fast Marching Methods can be used to provide fast ways to evaluate visibility terms.

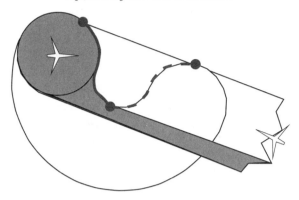

Fig. 20.18. Minimal unsafe set, shown as the shaded region, in the relative axis frame. The goal is to generate the relative distance between aircraft at which the aircraft may switch safely from mode 1 to mode 2. Aircraft must not switch from q_1 to q_2 to the right of the dashed line; aircraft must switch to the right of the solid line.

20.6.1 Profile visibility

As will be discussed in the next chapter on process simulation in semi-conductor manufacturing, one factor which controls the etching and deposition process is whether or not an point on an evolving profile is visible to an oncoming beam. As example, imagine an etching beam which dislodges particles as it strikes the surface profile; these dislodged particles leave the interface and are then re-deposited somewhere else on the profile. Thus, the total flux of particles into any point on the front depends on computing the amount received from all other points on the front. There are various models which describe how these dislodged particles are ejected, including specular reflection (the angle of incidence equals the angle of refraction) and luminescent reflection (the angle at which a dislodged particle leaves the surface is uniformly distributed among all possible angles). In most cases, visibility comes into play; the amount of material at point A on the surface received from dislodged particles emanating from another point B on the surface is zero if there is no direct line of sight between the two points. This means that some other part of the profile blocks the incoming flux from point B (see Figure 20.19a).

In order to evaluate this effect, we require a rapid way to evaluate whether two points on a profile can "see" each other. A brute force approach, while easy to program, is costly. Suppose that there are N

points on the curve. By checking all segments between the two points, it costs $O(N)$ to determine if A can see B. Of course, clever programming can make this much faster, including quadtree representations and judicious choices of how to proceed through the list of segments.

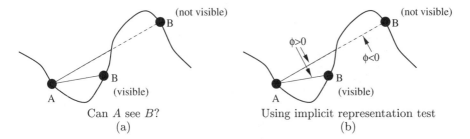

Fig. 20.19. Visibility test between A and B.

An efficient algorithm comes from exploiting our implicit representation of the interface. Suppose that the front is in fact represented as the level set of a function ϕ which is defined in a narrow band around the profile. Then we need only check to see if $\phi(x) > 0$ for all points x on the segment between A and B. If the level set function is always positive, then the two points are visible, since the front cannot intersect the segment. If we have chosen a motion so that the level set function is always close to the distance function, we can perform this test of whether $\phi(x) > 0$ along the segment easily as follows. First, pick the midpoint of the segment and evaluate ϕ by interpolation from the underlying grid; the value indicates the distance to the closest point on the front. Hence, one can move that distance along the line segment in both directions and then query again. Repeating this process until one either reaches both A and B (in which case the two points were mutually visible) or reaches a point where $\phi < 0$ (in which case the two points were not mutually visible) terminates the algorithm; see Figure 20.19b.

20.6.2 Visibility around obstacles

Another visibility problem may be summarized as follows. Given a collection of obstacles, find all points in a domain directly visible from a source at A. For a small number of obstacles, there are standard computer graphics techniques that typically perform comparison with each obstacle. As the number of obstacles increases, so does the complexity of the algorithm.

We can instead approach this problem from a partial differential equations perspective as follows. Given a source at A, imagine the solution to two separate Eikonal equations:

$$|\nabla T_{\text{no-obstacles}}| = 1 \qquad\qquad |\nabla T_{\text{obstacles}}| = F(x, y), \qquad\qquad (20.17)$$

where the right-hand side in the *obstacles* case is set to ∞ inside the obstacles and 1 otherwise. We solve each of these problems separately, using the Fast Marching Method, and then compare the solutions. If the obstacle blocks the source A from the point (x, y), then the first arrival time should be larger in the *obstacles* case. Thus, if $T_{\text{no-obstacles}}(x, y) = T_{\text{obstacles}}(x, y)$, then the point (x, y) is visible from the source; if, on the other hand, $T_{\text{no-obstacles}}(x, y) < T_{\text{obstacles}}(x, y)$, then the point is not visible. In practice, due to numerical error in the scheme, we can use one of two approaches. In the first approach, we use a threshold on the difference; that is, we check that

$$T_{\text{no-obstacles}}(x, y) + threshold < T_{\text{obstacles}}(x, y).$$

The threshold is chosen as a function of the mesh size h. In the second approach, we run the method twice; first with a given grid size and then with a grid size twice as fine, and this then indicates whether an obstacle is present.

We note two advantages with this approach. First, the computational speed is independent of the number of obstacles and second, the technique is unchanged in three dimensions and higher. In Figure 20.20, we show the shadow zone created around six obstacles for two different placements of the sources.

The technique may be easily extended to multiple sources by running each source separately and then comparing the separate arrival times (see Figure 20.21). For further discussion of visibility and related issues, see [239].

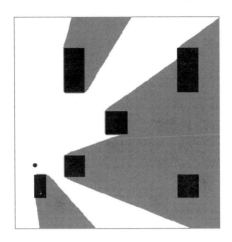

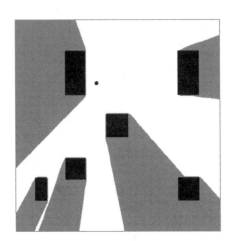

Fig. 20.20. Shadow zone created by obstacles.

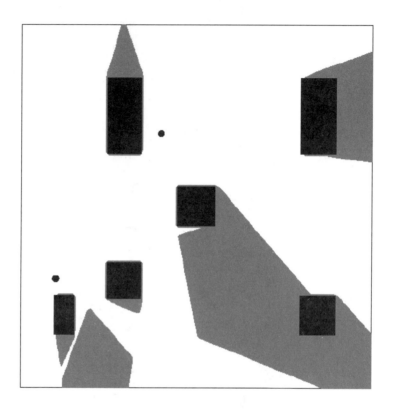

Fig. 20.21. Shadow zone for two sources.

21

Etching and Deposition in Microchip Fabrication

Outline: *We end with the application of Level Set and Fast Marching Methods to tracking interfaces in the microfabrication of electronic components. The goal is to follow the changing surface topography of a wafer as it is etched, layered, and shaped during the manufacturing process. These complex simulations rest on many of the previously discussed techniques, including Narrow Band Level Set methods, Fast Marching Methods for the Eikonal equation, construction of extension velocities, masking, discontinuous speed functions, visibility determinations, algorithms for subtle speed laws and second derivatives of curvature, and fast integral equation solvers. We present this final chapter in considerable detail, in order to give the flavor of how a complex interface application is assembled.*

21.1 Physical effects and background

The goal of numerical simulations in microfabrication is to model the process by which silicon devices are manufactured. Here, we briefly summarize some of the physical processes. First, a single crystal ingot of silicon is extracted from molten pure silicon. This silicon ingot is then sliced into several hundred thin wafers, each of which is then polished to a smooth finish. A thin layer of crystalline silicon is then oxidized, a light-sensitive "photoresist" that is sensitive to light is applied, and the wafer is then covered with a pattern mask that shields part of the photoresist. This pattern mask contains the layout of the circuit itself. Under exposure to a light or an electron beam, the exposed photoresist polymerizes and hardens, leaving an unexposed material that is then etched away in a dry etch process, revealing a bare silicon dioxide layer. Ionized impurity atoms such as boron, phosphorus, and argon are

313

then implanted into the pattern of the exposed silicon wafer, and silicon dioxide is deposited at reduced pressure in a plasma discharge from gas mixtures at a low temperature. Finally, thin films such as aluminum are deposited by processes such as plasma sputtering, and contacts to the electrical components and component interconnections are established. The result is a device that carries the desired electrical properties.

These processes produce considerable changes in the surface profile as it undergoes various effects of etching and deposition. This problem is known as the "surface topography problem" in microfabrication and is controlled by a large collection by physical factors, including the visibility of the etching/deposition source from each point of the evolving profile, surface diffusion along the front, complex flux laws that produce faceting, shocks and rarefactions, material-dependent discontinuous etch rates, and masking profiles.

The underlying physics and chemistry that contribute to the motion of the interface profile are very much areas of active research. Nonetheless, once empirical models are formulated, the problem ultimately becomes the familiar one of tracking an interface moving under a speed function F. This final chapter on Level Set and Fast Marching applications reviews some recent work on profile advancement for these problems. Simulations and text in this chapter are taken from Adalsteinsson and Sethian [3, 4, 5]; complete details may be found therein (see [241] for a review).

The underlying physical effects involved in etching, deposition, and lithography are quite complex; excellent overviews can be found in Scheckler [217], Scheckler, Toh, Hoffstetter, and Neureuther [218], and Toh and Neureuther [271]; see also Rey, Cheng, McVittie, and Saraswat [201], McVittie, Rey, Bariya, et al. [173], and Cale and Raupp [44, 45, 46]. The effects may be summarized briefly as follows:

- *Deposition:* Particles are deposited on the surface, which causes build-up in the profile. The particles may either isotropically condense from the surroundings (known as chemical or "wet" deposition), or be deposited from a source. In the latter case, particles leave the source and deposit on the surface; the main advantage of this approach is increased control over the directionality of surface deposition. The rate of deposition, which controls the growth of the layer, may depend on source masking, visibility effects between the source and surface point, angle-dependent flux distribution of source particles, and the angle of incidence of the particles relative to the surface normal direction. In addition, particles might not stick, but in fact be re-emitted back into the domain. This process is know as "re-emission" and the "sticking coefficient" between zero and one is the fraction of particles that stick. A sticking coefficient of unity means that all particles stick. Conversely, a low sticking coefficient means that particles may bounce many times before they eventually become fixed to the surface.

- *Etching:* Particles remove material from the evolving profile boundary. The material may be isotropically removed, known as chemical or "wet" etching, or chipped away through reactive ion etching, also known as "ion milling". Similar to deposition, the main advantage of reactive ion etching is enhanced directionality, which becomes increasingly important as device sizes decrease substantially and etching must proceed in vertical directions without affecting adjacent features. The total etch rate (see [248]) consists of an ion-assisted rate and a purely chemical etch rate due to etching by neutral radicals, which may still have a directional component. As in the above, the total etch rate due to wet and directional milling effects can depend on source masking, visibility effects between the source and surface point, angle-dependent flux distribution of source particles, and the angle of incidence of the particles relative to the surface normal direction. In addition, because of chemical reactions that take place on the surface, etching can cause surface particles to be ejected; this process is known as "re-deposition". The newly ejected particles are then deposited elsewhere on the front, depending on their angle and distribution.

- *Lithography:* The underlying material is treated by an electromagnetic wave that alters the resist property of the material. The aerial image is found, which then determines the amount of crosslinking at each point in the material. This produces the etch/resist rate at each point of the material. A profile is then etched into the material, where the speed of the profile in its normal direction at any point is given by the underlying etch rate.

In the rest of this section, we formalize the above. Define the coordinate system with the x and y axes lying in the plane, and z being the vertical axis. Consider a periodic initial profile $h(x, y)$, where h is the height of the surface above the x-y plane, as well as a source Z given as a surface above the profile; we write $Z(x, y)$ as the height of the source at (x, y). Define the source ray as the ray leaving the source and aimed toward the surface profile. Let ψ be the angle variation in the source ray away from the negative z axis; ψ runs from 0 to π, though it is physically unreasonable to have $\pi/2 < \phi < \pi$. Let γ be the angle between the projection of the source ray in the x, y plane and the positive x axis. Let n be the normal vector at a point x on the surface profile and θ be the angle between the normal and the source ray.

In Figure 21.1, these variables are indicated. Masks, which force flux rates to be zero, are indicated by heavy dark patches on the initial profile. At each point of the profile, a visibility indicator function $M_\Upsilon(x, x')$ is assigned which indicates whether the point x on the initial profile can be seen by the source point x'.

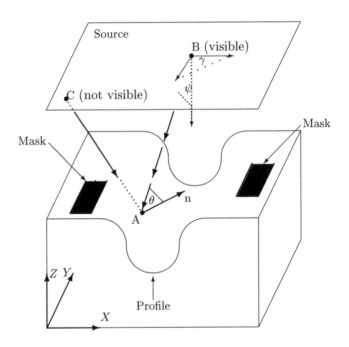

Fig. 21.1. Variables and setup.

21.2 Equations of motion for etching/deposition

The goal is to write the effects of deposition and etching on the speed F at a point x on the front. (We shall treat lithography separately, although it is amenable to the same unified level set approach). We refer the interested reader to a large collection of previous work on this topic, including [44, 45, 46, 112, 124, 147, 173, 201, 217, 218, 246, 248, 271].

21.2.1 Individual terms

21.2.1.1 Etching

We consider two separate types of etching:

- $F_{Isotropic}^{Etching}$: *Isotropic etching.* Uniform etching, also known as chemical or wet etching.
- $F_{Direct}^{Etching}$: *Direct etching.* Etching from an external source; this can be either a collection of point sources or an external stream coming from a particular direction. Visibility effects are included, and the flux strength can depend on both the solid angle from the emitting source and the angle between the profile normal and the incoming source direction. Etching can include highly sensitive dependence on angle such as in ion milling.

21.2.1.2 Deposition

We consider four separate types of deposition:

- $F_{Isotropic}^{Deposition}$: *Isotropic deposition.* Uniform deposition, also known as chemical or wet deposition.
- $F_{Direct}^{Deposition}$: *Direct deposition.* Deposition from an external source; this can be either a collection of point sources, or from an external stream coming from a particular direction. Visibility effects are included and the flux strength can depend on both the solid angle from the emitting source and the angle between the profile normal and the incoming source.
- $F_{Re-deposition}^{Deposition}$: *Re-deposition.* Particles that are expelled during the etching process. These particles then attach themselves to the profile at other locations. The strength and distribution of the re-deposition flux function can depend on such factors as the local angle. A re-deposition coefficient, $\beta_{Re-deposition}$ which can range from zero to unity represents the fraction of re-deposition that results from the

etching process. A value of $\beta_{Re-deposition} = 1$ means that nothing is re-deposited and everything sticks.

- $F_{Re-emission}^{Deposition}$: *Re-emission deposition*. Particles are deposited by direct deposition might not stick and are may be re-emitted into the domain. The amount of particles re-emitted depends on a sticking coefficient $\beta_{Re-emission}$. If $\beta_{Re-emission} = 1$, nothing is re-emitted.

In Figure 21.1, we generalize all of these effects as the "source." The plane source is shown in the figure may consist of locations which emit either unidirectional or point source contributions.

21.2.2 Assembling the terms

We may, somewhat abstractly, assemble the above terms into the single expression

$$F = F_{Isotropic}^{Etching} + F_{Direct}^{Etching} + F_{Isotropic}^{Deposition} + F_{Direct}^{Deposition} + F_{Re-deposition}^{Deposition} + F_{Re-emission}^{Deposition} \tag{21.1}$$

The two isotropic terms are evaluated at a point x by simply evaluating the strengths at that point. The two direct terms are evaluated at a point x on the profile by first computing the visibility to each point of the source, and then evaluating the flux function. These terms require computing an integral over the entire source. To compute the fifth term at a point x, we must consider the contributions of every point on the profile to check for re-deposition particles arising from the etching process; thus this term requires computing an integral over the profile itself. The sixth term, $F_{Re-emission}^{Deposition}$ is more problematic. Since every point on the front can act as a deposition source of re-emitted particles that do not stick, the total flux function deposition function comes from evaluating an integral equation along the entire profile.

In more detail, let Ω be the set of points on the evolving profile at time t, and let *Source* be the external source. Given two points x and x', let $\Upsilon(x, x')$ be one if the points are visible from one another and zero otherwise. Let r be the distance from x to x', let $\vec{n}$ be the unit normal vector at the point x, and finally, let $\vec{\alpha}$ be the unit vector at the point x' on the source pointing toward the point x on the profile. Then we may refine the above terms as:

$$
F = \begin{bmatrix}
\text{Flux}_{Isotropic}^{Etching} \\[2ex]
+ \\[2ex]
\int_{Source} \text{Flux}_{Direct}^{Etching}(r,\psi,\gamma,\theta,x)\Upsilon(x,x')(\vec{n}\cdot\vec{\alpha})dx' \\[2ex]
+ \\[2ex]
\text{Flux}_{Isotropic}^{Deposition} \\[2ex]
+ \\[2ex]
\int_{Source} \text{Flux}_{Direct}^{Deposition}(r,\psi,\gamma,\theta,x)\Upsilon(x,x')(\vec{n}\cdot\vec{\alpha})dx' \\[2ex]
+ \\[2ex]
\int_{\Omega}(1-\beta_{Re-deposition})\text{Flux}_{Re-deposition}^{Deposition}(r,\psi,\gamma,\theta,x)\Upsilon(x,x')(\vec{n}\cdot\vec{\alpha})dx' \\[2ex]
+ \\[2ex]
\int_{\Omega}(1-\beta_{Re-emission})\text{Flux}_{Re-emission}^{Deposition}(r,\psi,\gamma,\theta,x)\Upsilon(x,x')(\vec{n}\cdot\vec{\alpha})dx'
\end{bmatrix}
$$

$$(21.2)$$

The integrals are performed in a straightforward manner. The front is located by constructing the zero level set of ϕ; in two dimensions it is represented by a collection of line segments and in three dimensions by a collection of voxel elements; see [3, 4]. The centroid of each element is taken as the control point, and the individual flux terms are evaluated at each control point. In the case of the two isotropic terms, the flux is immediately found. In the case of the two integrals over sources, the source is suitably discretized and the contributions summed. In the fifth term, corresponding to re-deposition, the integral over the entire profile is calculated by computing the visibility to all other control points, and the corresponding re-deposition term is produced by the effect of direct deposition. Thus, the fifth term requires N^2 evaluations, where N is the number of control points which approximate the front.

21.2.3 Evaluation of the re-emission term

The sixth and last term is somewhat more time-consuming to evaluate, since it requires evaluation of the flux $Flux_{Re-emission}^{Deposition}$ from each point of the interface, each of which depends on the contribution from all other points. Thus, this is an integral equation which must be solved to produce the total deposition flux at any point. In the following discussion, we shall assume that the total deposition flux depends on deposition directly from the source, as well as additional deposition due to particles which do not stick and are re-emitted.[1] We shall call this flux $Flux_{Direct/Re-deposition}$ and solve for it.

We now introduce some additional notation; in this discussion, we derive the appropriate integral equation for two-dimensional problems; the derivation in three dimensions is similar. Let

- t_i be the coordinates of point number i, with an associated segment length l_i, $i = 1, \ldots, N$,
- r_{ij} be the distance between points i and j,
- θ_j^i be the angle between the normal to point i and the vector $t_j - t_i$,
- Υ_{ij} be the visibility factor, which is one if the points x_i and x_j can see each other, and zero otherwise,
- β_0 be the sticking coefficient for the particles coming directly from the source; $\beta_0 = 1.0$ means that all the particles stick,
- β be the sticking coefficient for secondary bounces,
- I_S^i be the incoming source strength at point i due to the above light source,
- I_R^i be the source strength radiated from point i, and
- I_S and I_R be the vectors $(I_S^0, \ldots, I_S^n)$ and $(I_R^0, \ldots, I_R^n)$.

The expression for the flux is an integral equation. Once the problem is discretized into a matrix relation, there are two numerical approaches to solving the equation. The first is to use a direct solver for the matrix equation. In two dimensions, this is practical; for large three dimensional problems this becomes impractical due to the computational labor. The second approach is to construct an iterative solution to the integral equation, based on a series expansion of the interaction matrix. We shall discuss both approaches; the text and explanation are taken without change from [5], and we refer the reader there for more details.

---

[1] Ion-induced re-deposition particles can be included as part of the re-emission process as well.

21.2.3.1 Direct solution of integral equation

The strategy, as discussed in [5], is to first work with the amount I_R radiated from each point of the surface. This depends on the amount received from the direct source plus the amount radiated from all other points on the surface, that is,

$$I_R^i = (1 - \beta_0)I_S^i + (1 - \beta) \sum_{j, j \neq i} I_R^j \frac{2\cos(\theta_i^j)\cos(\theta_j^i)}{\pi r_{ij}} \Upsilon_{ij} l_j. \tag{21.3}$$

In this equation, there are several things to point out. First, one stands at each point i and computes the contributions from the source, as well as from all other points on the front. Second, there are two contributing cosines in the expression. One is from the receiving point, which contains a cosine due the collected flux along the segment. The second cosine is due to the assumption of luminescent reflection: we assume that once a particle hits, it has an equal probability of heading off in any direction. Using the model of an assumed cosine distribution around the normal, this then generates the other cosine term.

Define the matrix Ω by

$$\Omega_{ij} = 2\frac{\cos(\theta_j^i)\cos(\theta_i^j)}{\pi r_{ij}}\Upsilon_{ij} l_j, \tag{21.4}$$

when $i \neq j$, and 0 if $i = j$.

Note that since $\cos(\theta_j^i) = n_i \cdot \frac{(t_j - t_i)}{|t_j - t_i|}$, Ω can be rewritten as

$$\Omega_{ij} = \frac{[n_i \cdot (t_j - t_i)][n_j \cdot (t_i - t_j)]}{\pi |t_j - t_i|^3}\Upsilon_{ij} l_j. \tag{21.5}$$

Thus,

$$I_R = (1 - \beta_0)I_S + (1 - \beta)\Omega I_R, \tag{21.6}$$

and we may express I_R in terms of I_S by

$$I_R = (1 - \beta_0)(I - (1 - \beta)\Omega)^{-1}I_S. \tag{21.7}$$

Thus, the received flux at the front at a point i is given by

$$Flux_{Direct/Re-deposition}^i = \beta_0 I_S^i + \beta \sum_{j, j \neq i} I_R^j \frac{2\cos(\theta_i^j)\cos(\theta_j^i)}{\pi r_{ij}}\Upsilon_{ij} l_j, \tag{21.8}$$

and, in vector form, we may rewrite this expression as

$$Flux_{Direct/Re-deposition} = \beta_0 I_S + \beta \Omega I_R. \tag{21.9}$$

The goal now is to eliminate the expression I_R in the above. We first rearrange Eqn. 21.6 to get

$$\Omega I_R = \frac{1}{(1 - \beta)} I_R - \frac{(1 - \beta_0)}{(1 - \beta)} I_S. \qquad (21.10)$$

Substitution into the flux equation (Eqn. 21.9 then gives

$$Flux_{Direct/Re-deposition} = \frac{(\beta_0 - \beta)}{(1 - \beta)} I_S + \frac{\beta}{(1 - \beta)} I_R. \qquad (21.11)$$

Finally, we may substitute the expression for I_R in terms of I_S to get

$$\frac{(\beta_0 - \beta)}{(1 - \beta)} I_S + \frac{\beta(1 - \beta_0)}{(1 - \beta)} (I - (1 - \beta)\Omega)^{-1} I_S. \qquad (21.12)$$

Note the important special case when $\beta_0 = \beta$ (that is, the sticking coefficient is the same for all bounces), in which case the equation becomes

$$Flux_{Direct/Re-deposition} = \beta(I - (1 - \beta)\Omega)^{-1} I_S. \qquad (21.13)$$

We note that in these equations, Ω is non-symmetric,

$$\Psi_{ij} = \frac{n_i \cdot (t_j - t_i) n_j \cdot (t_i - t_j)}{\pi |t_j - t_i|^3} \Upsilon_{ij}, \qquad (21.14)$$

and letting L be the diagonal matrix with $L_{ii} = l_i$, $\Omega = \Psi L$, and Ψ is symmetric. The equation therefore becomes

$$\frac{(\beta - \beta_0)}{(1 - \beta)} I_S + \frac{\beta(1 - \beta_0)}{(1 - \beta)} (I - (1 - \beta)\Psi L)^{-1} I_S, \qquad (21.15)$$

$$\frac{(\beta - \beta_0)}{(1 - \beta)} I_S + \frac{\beta(1 - \beta_0)}{(1 - \beta)} L^{-1} (L^{-1} - (1 - \beta)\Psi)^{-1} I_S. \qquad (21.16)$$

This matrix is full and can be quite substantial if the front is complex. The symmetric solver in LinPack is used; approximations to this equation and faster summation techniques can also be considered, and are the subject of current work.

21.2.3.2 Iterative solution of integral equation

We now discuss a different approach, developed by Adalsteinsson and Sethian [5], which constructs an iterative solution to the integral equation. This consists of a series expansion in the interaction matrix. Suitably interpreted, it can be viewed as a "multi-bounce" model, in which the number of terms in the series expansion corresponds to the number of bounces that a particle can undergo before its effects are negligible.

This approach allows one to check the error remainder term in this iterative formulation to determine how many terms must be kept. Since most of the particles either stick or leave the domain after a reasonable number of bounces, this is an effective approach.

Following [5], begin by defining the reflected intensity $I_{R,k}$ after the k-th bounce, namely,

$$I_{R,1}^i = (1 - \beta_0)I_S^i, \qquad (21.17)$$

$$I_{R,k+1}^i = (1 - \beta) \sum_{j,j \neq i} I_{R,k}^j \frac{2 \cos(\theta_i^j) \cos(\theta_j^i)}{\pi r_{ij}} \Upsilon_{ij} l_j. \qquad (21.18)$$

In matrix form, this becomes

$$I_{R,0} = (1 - \beta_0)I_S, \qquad (21.19)$$

$$I_{R,k+1} = (1 - \beta)\Omega I_{R,k}, \qquad (21.20)$$

where Ω is defined as before. Now, define $I_{S,k}$ to be the portion that sticks at the k-th bounce. We then have that

$$I_{S,0} = \beta_0 I_S, \qquad (21.21)$$

$$I_{S,1} = \beta\Omega(1 - \beta_0)I_S, \qquad (21.22)$$

and, in general,

$$I_{S,k+1} = \frac{\beta}{1 - \beta} I_{R,k+1} = (1 - \beta)\Omega I_{S,k}. \qquad (21.23)$$

Therefore, by reaching back to the initial expression, we have

$$I_{S,k} = \beta(1 - \beta)^{k-1}(1 - \beta_0)\Omega^k I_S, \qquad (21.24)$$

and thus the total intensity after N applications is given by

$$I_N = \beta(1 - \beta_0) \left[\sum_{k=1}^{N} (1 - \beta)^{k-1}\Omega^k \right] I_S + \beta_0 I_S. \qquad (21.25)$$

Each application of the operator may be viewed as either an additional term in the expansion or an additional included bounce.

We note that there is a recurrence relation for I_N given by

$$
\begin{aligned}
I_{N+1} &= \beta(1-\beta_0)\left[\sum_{k=1}^{N+1}(1-\beta)^{k-1}\Omega^k\right]I_S + \beta_0 I_S \\
&= \beta(1-\beta_0)\left[(1-\beta)\Omega\sum_{k=1}^{N}(1-\beta)^{k-1}\Omega^k + \Omega\right]I_S + \beta_0 I_S \\
&= (1-\beta)\Omega(I_N - \beta_0 I_S) + \beta(1-\beta_0)\Omega I_S + \beta_0 I_S \\
&= (1-\beta)\Omega I_N + (\beta - \beta_0)\Omega I_S + \beta_0 I_S.
\end{aligned}
$$

By constructing the remainder term $I_{N+1} - I_N$, we can measure the convergence of the expansion and keep enough terms to bound the error below a user-specified tolerance.

21.3 Additional numerical issues

In order to use Level Set and Fast Marching Methods on these expressions, there are a few additional issues to tackle.

21.3.1 Visibility

In order to perform these integrations, we must evaluate the visibility between any two points. Here, we use the methodology described earlier: quick evaluations of visibility using the level set function ϕ. If ϕ is always positive between two points on the surface, then the two points are mutually visible.

21.3.2 Extension velocities

Both the total flux and the visibility have meaning only on the front itself. In this case, the front extension methodology described earlier is required in order to build values for the speed function throughout the narrow band.

21.3.3 Surface diffusion

An additional physical effect comes from surface diffusion. The surface profile is altered by surface diffusion along the front; this corresponds to motion by the second derivative of curvature, which is itself a highly delicate calculation.

Thus, we need to add an additional term of the form

$$F = 1 + \epsilon \kappa_{\alpha\alpha}, \tag{21.26}$$

where α is an arc-length parameterization. This term models the effects of surface diffusion, which relates to the motion of metal boundaries; see [43, 121] for experimental evidence.

The problem is delicate because Eqn. 21.26 is a time-dependent fourth order partial differential equation, and the presence of the fourth derivative requires an exceedingly small time step for stability in an explicit scheme; the linear fourth order heat equation has a stability time step requirement of the form $O(\Delta t/\Delta h^4)$. We have seen such an equation earlier in the chapter on geometric flows. We point out that such schemes can in fact be made implicit to allow a larger time step; see [62] for further discussion.

21.4 Two-dimensional results

21.4.1 Etching/deposition

Figure 21.2 shows a deposition source above a trench, with deposition material emitted from a line source represented as a solid line above the trench. The deposition rate is the same in all directions, and shadowing effects are considered. Figure 21.2(a) shows results for 40 computational cells across the width of the computational region (between the two vertical dashed lines), Figure 21.2(b) has 80 cells, and Figure 21.2(c) has 160 cells. The time step for all three calculations is $\Delta t = .00625$. The calculations are performed with a narrow band tube width of 6 cells on either side of the front. There is little change between the calculation with 80 cells and the one with 160 cells, indicating a converged solution. As the walls pinch toward each other, the visible angle decreases and the speed diminishes.

Next, consider directional etching into a trench/cavity. Material emits from the line source at an angle of 30 degrees from the vertical. Figure 21.3 shows results at various times, starting with the initial state. Again, there is no yield variation in the etch rate due to the angle of incidence with the normal; the speed of the profile in the normal direction is just the projection of the directional etch rate in that normal direction. Because of the effects of shadowing, as the profile evolves it aligns itself along the incoming unidirectional etching stream.

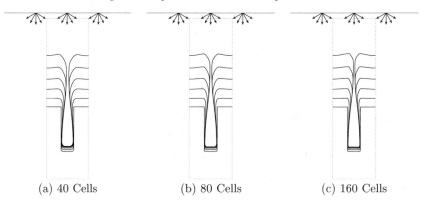

(a) 40 Cells   (b) 80 Cells   (c) 160 Cells

Fig. 21.2. Source deposition into trench.

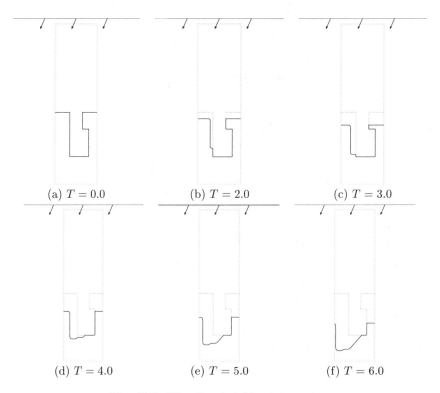

(a) $T = 0.0$ (b) $T = 2.0$ (c) $T = 3.0$

(d) $T = 4.0$ (e) $T = 5.0$ (f) $T = 6.0$

Fig. 21.3. Directional etching into cavity.

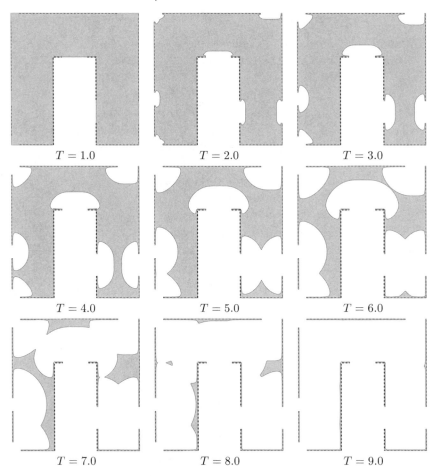

$T = 1.0$ $T = 2.0$ $T = 3.0$

$T = 4.0$ $T = 5.0$ $T = 6.0$

$T = 7.0$ $T = 8.0$ $T = 9.0$

Fig. 21.4. Chemical etching into multiply masked region.

21.4.2 Masking

The effect of masking is studied using a square with masks covering segments of the boundary. Imagine the square surrounded by an etching substance. In Figure 21.4 the etch eats into the non-masked walls. The resulting front moves into the region and reconnects with other parts of the advancing front.

21.4.3 Ion milling: non-convex sputter laws

A more sophisticated set of examples arises in simulations (for example, of ion milling) in which the normal speed of the profile depends on the angle of incidence between the surface normal and the incoming beam. This yield function is often empirically fit from experiment, and it has been observed to cause such effects as faceting at corners (see Leon et al. [147] and Katardjiev, Carter, and Nobes [124]). As shown in [3, 4], such yield functions can often give rise to non-convex Hamiltonians, in which case alternative schemes must be used.

As an example, consider an etching beam coming down in the vertical direction. In the cases under study here, the angle θ shown in Figure 21.1 refers to the angle between the surface normal and the positive vertical. For this set of calculations, in order to focus on the geometry of sputter effects on shocks/rarefaction fan development, visibility effects are ignored. Let $F(\theta)$ be the speed of the front in direction normal to the surface, and consider three different speed functions:

- $F(\theta) = 1$,
- $F(\theta) = \cos(\theta)$,
- $F(\theta) = [1 + 4\sin^2(\theta)]\cos(\theta)$.

The first case corresponds to isotropic etching. We shall now show that the third case leads to a non-convex Hamiltonian. We have

$$\phi_t + F|\nabla\phi| = \phi_t + [(1+A)\cos\theta - A\cos^3\theta]|\nabla\phi|. \qquad (21.27)$$

Noting that $\cos\theta = \frac{\phi_y}{|\nabla\phi|}$, some manipulation produces

$$\phi_t + H(\phi_x, \phi_y) = 0, \qquad (21.28)$$

where the Hamiltonian H is now $H = (1+A)\phi_y - A\frac{\phi_y^3}{|\nabla\phi|^2}$. It is easy to check that this Hamiltonian is non-convex for $A > 0$.

Thus, in the case of this non-convex Hamiltonian, appropriate non-convex schemes of the sort given earlier are required. Figure 21.5 shows the results of applying both the convex and the non-convex schemes. In column A, the effects of purely isotropic motion are shown; thus the yield function is $F = 1$. Located above the yield graph are the motions of a downward square wave under etching. The top row is calculated using the convex scheme, and the second row using the non-convex scheme. In column B, the effects of directional motion are shown; thus the yield function is $F = \cos(\theta)$. In this case, the horizontal components on the

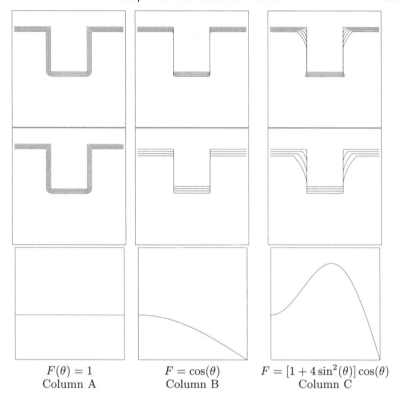

$$F(\theta) = 1 \qquad F = \cos(\theta) \qquad F = [1 + 4\sin^2(\theta)]\cos(\theta)$$
Column A   Column B   Column C

Fig. 21.5. Ion milling under various yield functions. Top row: convex scheme; Middle row: non-convex scheme; Bottom row: yield curves.

profile do not move, and vertical components move with unit speed. Column C shows the effects of the yield function $F = [1 + 4\sin^2(\theta)]\cos(\theta)$.

The results of these calculations are given in Figure 21.5. The results show that the effects of angle-dependent yield functions are pronounced. In column A, the isotropic rate produces smooth corners, correctly building the necessary rarefaction fans in outward corners and entropy-satisfying shocks in inward corners, as discussed and analyzed in [222, 225]. In column B, the directional rate causes the front to be essentially translated upward, with minimal rounding of the corners. In column C, the yield function results in faceting of inward corners where shocks form together with smooth regions. We note that the application of the convex scheme to the non-convex Hamiltonian in column C leads to incorrect results, whereas application of the non-convex scheme produces the expected answer.

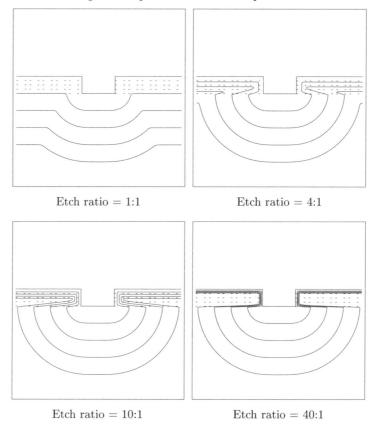

Etch ratio = 1:1   Etch ratio = 4:1

Etch ratio = 10:1   Etch ratio = 40:1

Fig. 21.6. Etch ratio = Bottom material rate to top material rate.

21.4.4 Discontinuous etch rates

Next, the effects of etching through different materials are studied. Here, the etch rates are discontinuous, and sharp corners develop in the propagating profile. The results are shown in Figure 21.6. A top material masks a lower material. The profile etches through the lower material first and underneath the upper material. The profile depends on the ratio of the etch rates. In Figure 21.6(a), the two materials have the same etch rate. In this case, the front simply propagates in its normal direction with unit speed, regardless of which material it is passing through. In Figure 21.6(b), the bottom material etches four times faster than the top; in Figure 21.6(c), the ratio is 10 to 1. Finally, in Figure 21.6(d), the ratio is 40 to 1, and hence the top material acts almost like a mask.

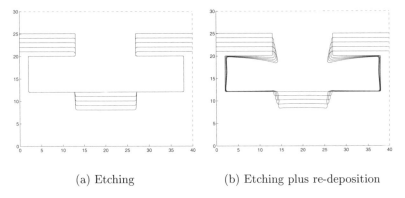

(a) Etching   (b) Etching plus re-deposition

Fig. 21.7. Effect of re-Deposition on unidirectional etching process.

21.4.5 Test cases: re-emission/re-deposition simulations

We begin with straightforward test examples. In Figure 21.7, etching occurs under the influence of a unidirectional vertical etching beam. In Figure 21.7a, there is no re-deposition. Conversely, in Figure 21.7b, an amount of material equal to the amount etched is re-emitted as a point source at the etching point and re-deposited elsewhere on the front: constant volume is maintained between the etching and re-emission processes.

Next, we consider a pure deposition process, in which the sticking coefficient is varied. In Figure 21.8a, a unidirectional deposition beam enters from the vertical, and all of the material sticks (sticking coefficient 1.0). In Figures 21.8b (21.8c), the sticking coefficient is $\beta = 0.5$ ($\beta = 0.2$). For sticking coefficient values less than unity, the matrix equation is solved. As the sticking coefficient decreases, the deposition layer becomes more uniformly distributed. We now turn to more complex simulations designed to demonstrate various physical effects.

21.4.6 Parameter studies of etching and deposition

We begin with a two-dimensional parameter study of the simultaneous effects of etching and deposition, ignoring the effects of re-deposition and re-emission. We use the speed function[2]

$$F = (1 - \alpha)F_{etch} + \alpha F_{Deposition}, \qquad (21.29)$$

---

[2] This form for the ion milling term was suggested by J. Rey.

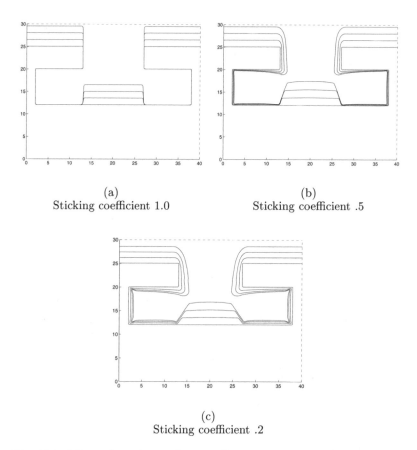

(a)
Sticking coefficient 1.0

(b)
Sticking coefficient .5

(c)
Sticking coefficient .2

Fig. 21.8. Effect of varying sticking coefficient on deposition/re-deposition.

where

$$F_{etch} = (5.2249 \cos \theta - 5.5914 \cos^2 \theta + 1.3665 \cos^4 \theta), \qquad (21.30)$$

$$F_{Deposition} = \beta F_{Isotropic} + (1 - \beta) F_{Source}. \qquad (21.31)$$

As shown in [3, 4], ion-milling terms of this form yield non-convex Hamiltonian-Jacobi equations and must be computed using appropriate upwind schemes. Visibility effects are considered in all terms except isotropic deposition. The results of varying α and β between 0 and 1 are shown in Figure 21.9.

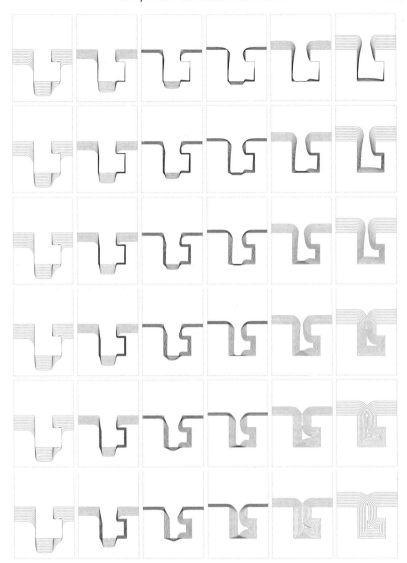

$$F = (1 - \alpha)F_{etch} + \alpha F_{Deposition}$$

$$F_{etch} = (5.2249 \cos\theta - 5.5914 \cos^2\theta + 1.3665 \cos^4\theta) \cos\theta$$

$$F_{Deposition} = \beta F_{Isotropic} + (1 - \beta)F_{Source}$$

α increases from left to right
β increases from top to bottom

Fig. 21.9. Simultaneous etching and deposition.

21.4.7 Trench depth on re-emission profiles

Next, we study the relationship between body geometry and re-emission profiles. In Figure 21.10, we show a 3 × 3 matrix displaying the interplay between body geometry and various values for the sticking coefficient. We assume a unidirectional deposition beam. A sticking coefficient of β means that $1 - \beta$ of the material is not deposited, but instead is re-emitted as a point source.

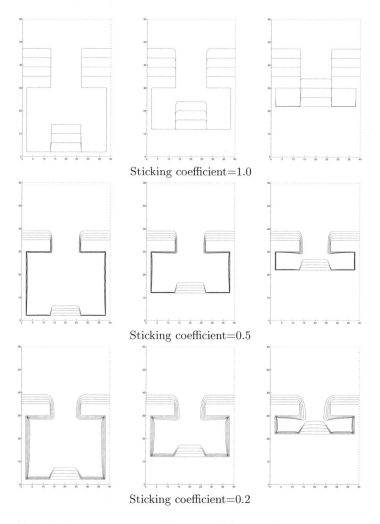

Fig. 21.10. Body geometry vs. sticking coefficient: unidirectional deposition.

We repeat the study in Figure 21.11, only this time we assume deposition from a line source above the trench. In both cases, as the depth of the cavity decreases, more re-emission is felt from the bottom of the cavity and the deposition spread is more uniform. Furthermore, in both cases we observe a slight lagging of the front in corners; this is due to leakage in our discretization of the integral for the front. The problem can be corrected with a non-uniform discretization scheme, which is discussed elsewhere; see [178].

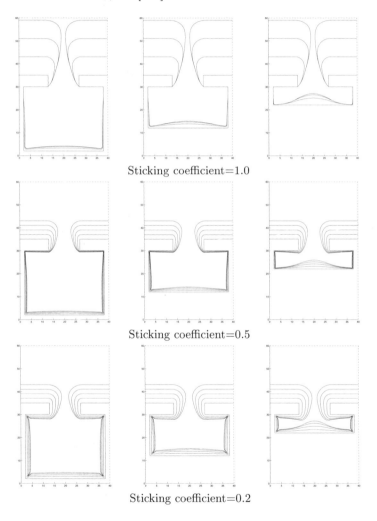

Sticking coefficient=1.0

Sticking coefficient=0.5

Sticking coefficient=0.2

Fig. 21.11. Body geometry vs. sticking coefficient: line source deposition.

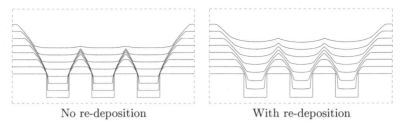

No re-deposition   With re-deposition

Fig. 21.12. Combination of ion milling, direct deposition and conformal deposition.

21.4.8 Multiple effects

An important simulation is obtained by considering a periodic sequence of structures under the multiple effects of ion-milling, isotropic deposition, and re-deposition. Here, the goal is to model the faceting that occurs due to the non-convexity of the ion-milling term (see [3, 4]), as well as the role of re-deposition in rounding sharp corners as a function of the re-emission coefficient. A combination of ion-milling, deposition, and ion-induced sputtered re-deposition is shown in Figure 21.12, together with conformal deposition and direct deposition. On the left, the ion-induced sputter re-deposition is set to zero. On the right the etched material is re-emitted, producing considerable rounding of the sharp corners.

Next, we consider a complex speed function,[3] which consists of a sensitive angle dependent speed law. A plot of the speed as a function of θ shows that for some values of θ, deposition dominates over etching, while for other values, etching is the dominant effect. We add the further restriction that the initial structure is impenetrable and thus cannot be etched. In Figure 21.13, we show the effect of the speed law

$$F = (4.385 - 5.7 \cos\theta + 1.425 \cos^3 \theta) \cos\theta \qquad (21.32)$$

on a periodic structure. We observe that the impenetrability of the material forces the selection of two critical angles, as seen in the sharp angles at the protruding corners of the structure.

21.4.9 Thin films and nanolayers

Next, we consider a problem in which several effects are combined. We imagine an initial block in which a mask covers a substrate, and we

---

[3] This example was suggested by J. Rey.

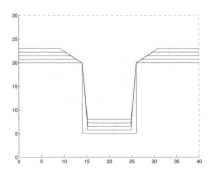

Fig. 21.13. Impenetrable periodic block under simultaneous etching and deposition.

envision simultaneous etch and deposition processes. We imagine that one material (which will be shown as light gray) is isotropically deposited on both the mask (shown in dark gray) and the substrate (shown in black). At the same time that this material is being deposited, it is being etched under an ion-milling/sputter law such that the etch rate in the substrate is twice as fast as the etch rate in the mask. Thus, we have

$$F = F_{IsotropicDeposition} + F_{SputterEtching} \qquad (21.33)$$

$$F_{IsotropicDeposition} = .5 \qquad F_{SputterEtching} = Factor(x) * \cos\theta$$
$$(21.34)$$

- $Factor(x)$= 1.0 if in dark gray material (Mask),
- $Factor(x)$= 2.0 if in black material (Substrate).

In Figure 21.14, we show the sequence of profile evolution under these effects. We note the development of the thin nanolayer which covers the side walls but is fully etched away along the top and the bottom. We also note the existence of evolving points where several fronts touch. We stress that the grid used for this calculation is significantly larger than the size of the nano-layer. Thus, our algorithms provide for significant sub-grid resolution without resorting to adaptive mesh technology; see [236] for details of this technology.

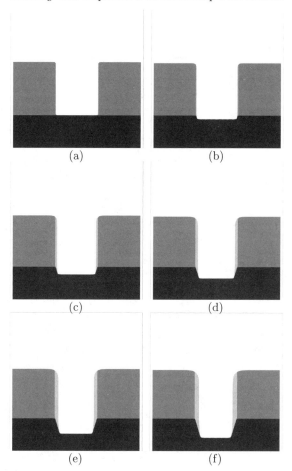

Fig. 21.14. Combination of isotropic deposition of thin layer and convex sputter etching of materials: time sequence.

$$F = F_{IsotropicDeposition} + F_{SputterEtching} \qquad (21.35)$$

$$F_{IsotropicDeposition} = .5 \qquad F_{SputterEtching} = Factor(x) * \cos\theta \quad (21.36)$$

$Factor(x)$= 1.0 if in mask (dark gray) / $Factor(x)$= 2.0 if in substrate(black)

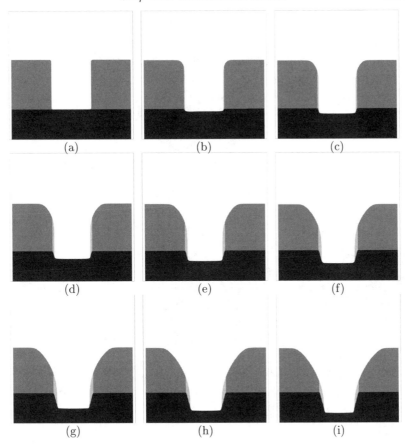

Fig. 21.15. Combination of isotropic deposition of thin layer and non-convex sputter etching of materials: time sequence.

$$F = F_{IsotropicDeposition} + F_{SputterEtching} \qquad (21.37)$$

$$F_{IsotropicDeposition} = .5 \qquad F_{SputterEtching} = Factor(x)*(1.+4\sin^2\theta)\cos\theta$$
$$(21.38)$$

$Factor(x)$= 1.0 if in mask (dark gray)/ $Factor(x)$= 2.0 if in substrate(black)

We repeat the calculation in Figure 21.15, this time using an ion-milling sputter etching speed law which promotes faceting due to the presence of non-convex Hamiltonians (see [3, 4]). Here, we note the rounding of the side wall layers, as well as the existence of multiple fronts and thin layer structures.

21.4.10 Surface diffusion

Two types of surface diffusion can play important roles in coverage and deposition layers: bulk diffusion, which is the global macro-motion of the material within the deposited layer, and surface diffusion, which relates to the motion of metal boundaries. Here, we examine the effects of surface diffusion on the shape of the deposition layer.[4]

Cale and Jain [43, 121] have performed carefully fit numerical experiments to match experimental evidence of surface diffusion effects of aluminum-(1.5%)copper films. They propose (see Cale and Jain [43], and Cale and Raupp [44, 45, 46]) a model of the form

$$\eta(s) - R(s) + \text{constant}\frac{\partial^2 \kappa}{\partial s^2} = 0, \qquad (21.39)$$

where $\eta(s)$ is the ballistic flux of atoms arriving at the surface position s, $R(s)$ is the rate of incorporation of atoms into the solid film, and κ is the signed curvature. We refer the reader to [44, 45, 46] for a detailed discussion of transport equations and related terms. We begin in Figure 21.16a by showing the effects of surface diffusion on a model problem of isotropic deposition, that is, we examine a speed function $F = 1 + \epsilon\kappa_{\alpha\alpha}$ for varying values of ϵ.

We next turn to a more realistic case and consider a speed function which contains isotropic deposition and an ion-milling non-convex etch function together with surface diffusion. In Figure 21.16b we consider the speed function

$$F = 4 - (1 + 4\sin^2\theta) + \epsilon\kappa_{\alpha\alpha}. \qquad (21.40)$$

---

[4] We thank T. Cale and J. Rey for illuminating conversations about the role and effects of surface diffusion.

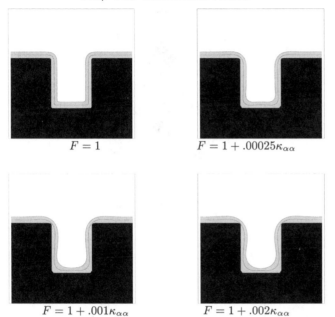

(a) Effects of surface diffusion on isotropic deposition: $F = 1 - \epsilon\kappa_{\alpha\alpha}$

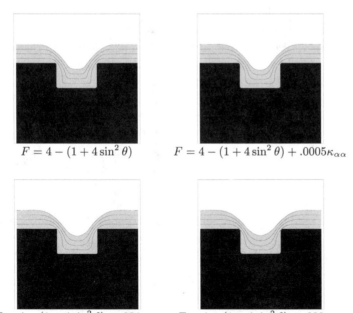

(b) Effects of surface diffusion on deposition plus ion milling: $F = 4 - (1 + 4\sin^2\theta) + \epsilon\kappa_{\alpha\alpha}$

Fig. 21.16. Effects of surface diffusion.

Fig. 21.17. Lithographic development on $50 \times 50 \times 50$ grid.

21.5 Three-dimensional simulations

21.5.1 Photolithography development

We begin three-dimensional simulations with a problem in photolithography. Once the electromagnetic and optical simulations are performed, the problem of photolithography development reduces to that of following an initially plane interface propagating downward in three dimensions. The speed in the normal direction is given as a supplied rate function at each point. The speed $F = F(x, y, z)$ depends only on position; however, it may change extremely rapidly. The goal in lithography development is to track this evolving front. In order to develop realistic structures in three-dimensional development profiles, a grid of size $300 \times 300 \times 100$ is not unreasonable. A fast algorithm is of considerable value in the development step.

In Sethian [233], the Fast Marching Method has been applied to lithography development. As a warm-up test example, Figure 21.17 shows the evolution until $T = 10$ of a flat profile at height $z = 1$ in the unit cube centered at $(.5, .5, .5)$ under a model Gaussian rate function given by

$$F(x, y, z) = e^{-64(r^2)}(\cos^2(12z) + .01), \qquad (21.41)$$

where $r = \sqrt{(x - .5)^2 + (y - .5)^2}$. This rate function F models the effect of standing waves which change the resist properties of the material and cause sharp undulations and turns in the evolving profile.

t]

Grid Size	Time to load rate file	Time to propagate front	Total time
50x50x50	0.1 secs	0.5 secs	0.6 secs
100x100x100	1.2 secs	5.1 secs	6.3 secs
150x150x150	3.9 secs	20.0 secs	23.9 secs
200x200x200	9.0 secs	55.0 secs	64 secs

Fig. 21.18. Timings for development to $T = 10$: Sparc 10.

Grid size	50^3	100^3	150^3	200^3	Exact
Breakthrough time T_{final}	9.8090801	10.33976	10.36608	10.36669	10.36761
Relative Error $\left(\dfrac{|\text{Computed} - \text{Exact}|}{\text{Exact}} \right)$	.05387	.00268	.00014	.00008	0.0

Fig. 21.19. Accuracy of calculation of breakthrough time.

Figure 21.18 shows timings for our Gaussian speed function. Note that loading the rate file is a significant part of the total compute time.

In Figure 21.19 accuracy is evaluated using the breakthrough time T_{final} when the profile reaches the bottom center point. The exact answer (computed using fourth order Runge-Kutta) is $T_{final} = 10.36761$.

To show a more realistic example of lithography development, a rate function calculated using the three-dimensional exposure and post-exposure bake modules of TMA's Depict 4.0 [269] is coupled to the Fast Marching Method. Figure 21.20(a) shows the top view of a mask placed on the board. The dark areas correspond to areas that are exposed to light. The standing waves in the etching profile are due to factors such as the reflectivity of the surface. In Figure 21.20(b) a view of the developed profile is shown from underneath; the etching of the holes and the presence of standing waves can be seen easily. For further results, see [234].

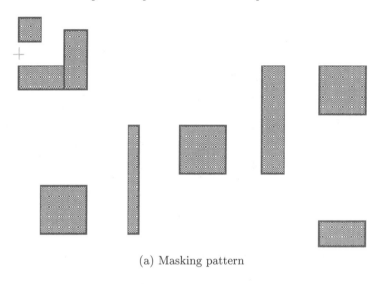

(a) Masking pattern

(b) Lithographic development: View from below

Fig. 21.20. Lithographic development using Fast Marching Method.

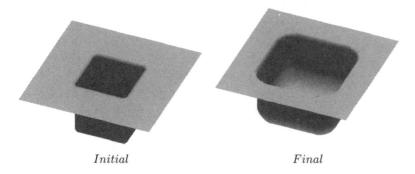

Initial   *Final*

Fig. 21.21. Isotropic etching into a hole.

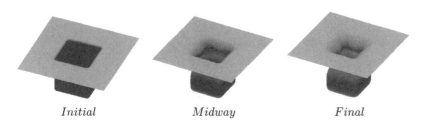

Initial   *Midway*   *Final*

Fig. 21.22. Source deposition into a hole.

21.5.2 Etching and deposition

We begin with a straightforward calculation of isotropic etching into a hole, taken from [4]. In Figure 21.21 we show a square hole from which a material is being isotropically etched, corresponding to a simple speed function of $F = -1$. As expected, the sides of the cavity are cleanly etched away, leaving smoothed, rounded walls.

We follow with a calculation of source deposition from a plate located above the hole. The effects of visibility and shading are included. Along the entire plate, deposition material is emitted uniformly in each direction. In Figure 21.22, we show two three-dimensional time plots of the evolving profile. The trench begins to pinch off due to the effects of visibility and a bulb-shaped profile evolves.

Next, Figure 21.23 shows the effects of unidirectional etching under a bridge structure. The bridge initially has a thin curtain stretched underneath it; the thickness of the curtain is smallest at the middle. Here, the pillars shadow the profile, and their effect can be seen on the flat part of the surface as the bridge is etched away.

We end the basic calculations section with the modeling (Figure 21.24)

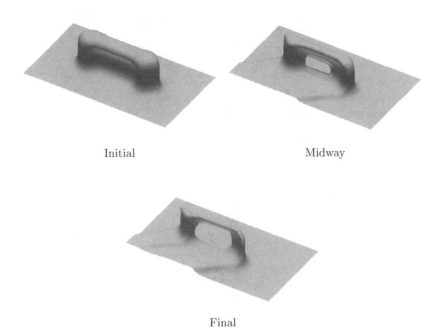

Initial   Midway

Final

Fig. 21.23. Unidirectional etching of bridge.

of the effect of non-convex sputter etch/ion milling of a saddle surface. The non-convex speed law $F = (1 + 4\sin^2(\theta))\cos\theta$ causes faceting of sharp corners and rounded polishing; for details of this effect, see [4].

21.5.3 Complex simulations

Next, we include an example of three-dimensional effects of re-deposition. The initial shape is a double-L, and we consider a combination of two cosine flux deposition sources. That is, the initial flux at each point is given by

$$Flux(x) = \cos^5(\theta_1)\cos(\theta_2) + \cos(\theta_1)\cos(\theta_2); \qquad (21.42)$$

in addition, the second deposition term is given a sticking coefficient of 0.1, thus we also consider the effects of re-deposition. Here, θ_1 is the angle that the vector v from x to y makes with the normal at x, and θ_2 is the angle that the vector v makes with the vertical. The results

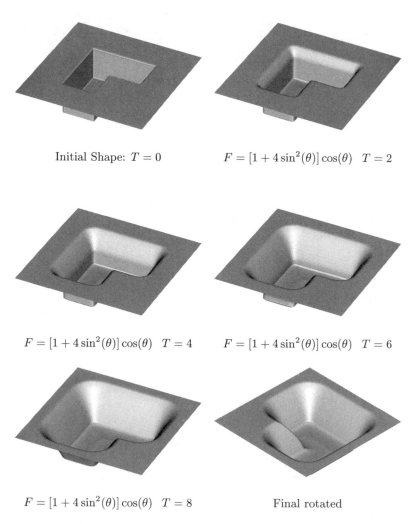

Initial Shape: $T = 0$ $F = [1 + 4\sin^2(\theta)]\cos(\theta)$ $T = 2$

$F = [1 + 4\sin^2(\theta)]\cos(\theta)$ $T = 4$ $F = [1 + 4\sin^2(\theta)]\cos(\theta)$ $T = 6$

$F = [1 + 4\sin^2(\theta)]\cos(\theta)$ $T = 8$ Final rotated

Fig. 21.24. Downward saddle under sputter etch.

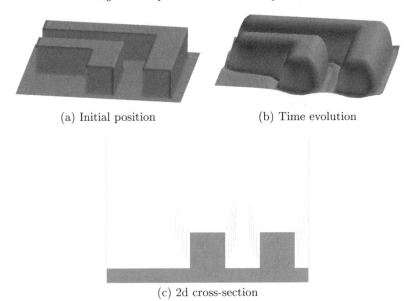

(a) Initial position   (b) Time evolution

(c) 2d cross-section

Fig. 21.25. Three-dimensional evolution under cosine source distribution with sticking coefficient 0.1.

are shown after some time evolution in Fig. 21.25b; a two-dimensional cross-sectional cut is shown in Figure 21.25c.

21.6 Timings

The computational labor required in these calculations depends on the grid resolution required to represent the front and the complexity of the physical effects under consideration. Figure 21.26 and Figure 21.27 give rough timings for various sizes and physical complexities for a Sun Ultra. The lithography timings were computed using the Fast Marching Method given in [233].

Test	50 by 50			100 by 100		
	Run time	Steps	time/step	Run time	Steps	time/step
Lithography (Fast Marching)	6.9ms	NA	NA	26ms	NA	NA
Isotropic (Narrow Band)	82ms	24	34ms	0.4s	49	8ms
Unidirectional (with visibility)	0.4s	17	23ms	2.3s	34	70ms
Etching and re-deposition	1.7s	25	68ms	14s	51	0.3s
Deposition and re-deposition (Iterative model)	1.1s	17	65ms	12s	39	0.3s

Fig. 21.26. Two-dimensional timings.

Test	40 by 40 by 40			80 by 80 by 80		
	Run time	Steps	time/step	Run time	Steps	time/step
Lithography (Fast Marching)	0.16s	NA	NA	2.1s	NA	NA
Isotropic (Narrow Band)	1.3s	8	0.16s	13.6s	24	0.6s
Unidirectional (with visibility)	16.7s	24	0.7s	270s	47	5.7s
Etching and re-deposition	224s	12	19s	260m	25	10m
Deposition and re-deposition (Iterative model)	265s	11	24s	290m	23	12.6m

Fig. 21.27. Three-dimensional timings.

21.7 Validation with experimental results

We end both this chapter and the book with a collection of applications of the Level Set/Fast Marching methodology comparing simulations with experiment analyzing various aspects of surface thin film physics. All the simulations in this section are performed using TERRAIN;[5] a commercial version of these techniques built by Technology Modeling Associates and specifically designed for process simulation. For further details about this code and its capabilities, see [270].

---

[5] We thank Juan Rey, Brian Li, and Jiangwei Li for providing these results.

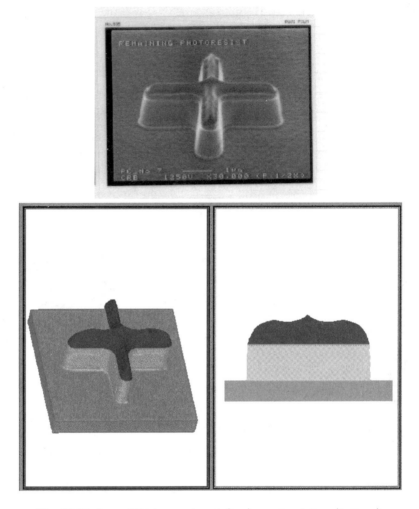

Fig. 21.28. Ion-milling: experiment (top) vs. simulation (bottom).

21.7.1 Ion milling

We begin with a comparison with experiment of an ion-milling process. Figure 21.28 shows an experiment on the top and a simulation at the bottom. We note that both the simulation and the experiment show the crossing non-convex curves on top of the structures, the sharp points, and the sloping sides.

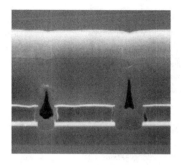 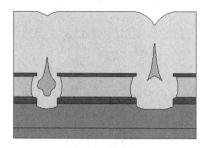

Fig. 21.29. PECVD, small-scale structure: experiment (left) vs. simulation (right).

 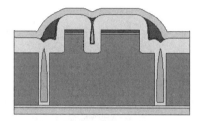

Fig. 21.30. PECVD, small-scale structure: experiment (left) vs. simulation (right).

21.7.2 Plasma-enhanced chemical vapor deposition

Next, we show comparison with experiment of two plasma-enhanced chemical vapor deposition (PECVD) simulations. We show a series of experiments. First, two smaller structure calculations are used to verify the ability to match experiment. Figures 21.29 and 21.30 show these results. Figures 21.31 and 21.32 show more simulations for more complex structures.

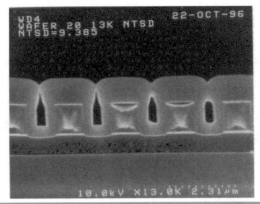

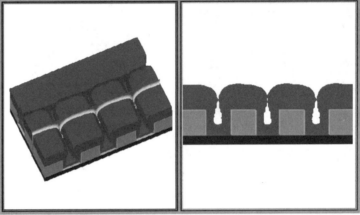

Fig. 21.31. PECVD: experiment (top) vs. simulation (bottom).

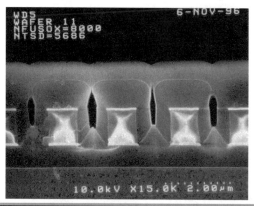

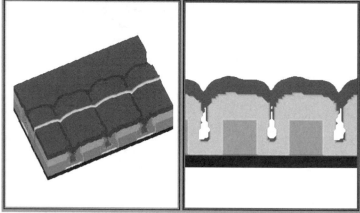

Fig. 21.32. PECVD: experiment (top) vs. simulation (bottom).

21.7.3 Spin-on-glass

Next, in Figure 21.33, we show a spin-on-glass (SOG) simulation; in which the spin deposition is shown in sequence on top of a complex structure.

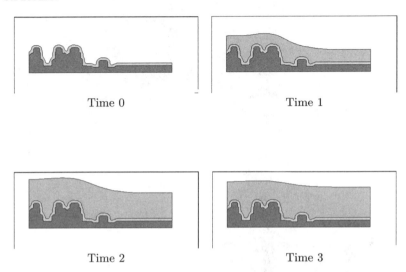

Time 0   Time 1

Time 2   Time 3

Fig. 21.33. Spin-on-Glass: simulation time sequence.

21.7.4 SRAM simulations

Finally, we show SRAM comparisons between experiment and simulations for both small structures (Figure 21.34) and large structures (Figure 21.35). Each figure shows the original layout together with the actual pattern printed through photolithography, followed by the sequential processing steps.

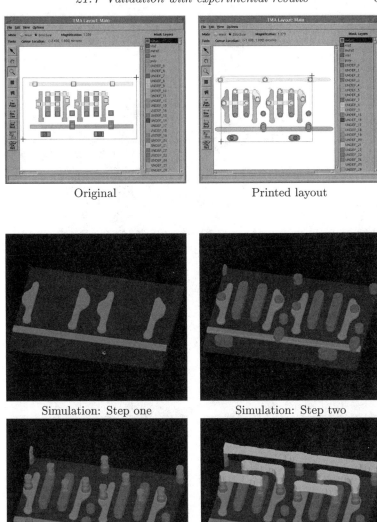

Original   Printed layout

Simulation: Step one   Simulation: Step two

Simulation: Step three   Simulation: Step four

Fig. 21.34. SRAM simulation: Experiment and simulation.

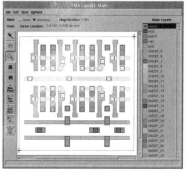
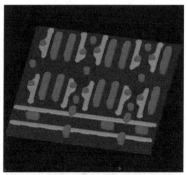

Original   Printed layout

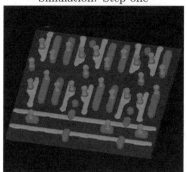
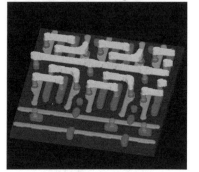

Simulation: Step one   Simulation: Step two

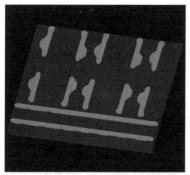

Simulation: Step three   Simulation: Step four

Fig. 21.35. SRAM simulation: Experiment and simulation.

22

Summary/New Areas/Future Work

To expand on a sentiment expressed in the first edition of this book, the range of Level Set Methods and now Fast Marching Methods, extends far beyond the work covered here. In particular, the algorithms have now reached a level of maturity that one need not rethink the formulation for each new topic. As a general guideline, suppose one has an equation of motion for an evolving interface. The underlying physics/chemistry which drives the interface is solved using either a finite difference or finite element grid. This produces a velocity on the interface. One then uses the extension velocity methodology, based on the Fast Marching Method and presented earlier, to build the velocities for each neighboring level set. This provides an update law for the Narrow Band Level Set Method at all nearby grid points. These values are updated at grid points, along with the grid values of the appropriate physical quantities, and the entire problem is advanced one time step. While faster, more accurate, and sometimes more artful versions can be devised for particular physical problems, this approach is robust, reliable, and straightforward.

As brief closing references, we direct the reader to some additional topics. On the theoretical side, considerable analysis of level set methods has been performed in recent years; see, for example, Brakke [35], Ecker and Huisken [80], Evans and Spruck [88, 89, 90, 91], Chen, Giga, Goto, and Ishii [57, 101, 102], and Ambrosio and Soner [11]. These works have concentrated on many aspects, including questions of existence and uniqueness, pathological cases, extensions of these ideas to fronts of co-dimension greater than one (such as evolving curves in three dimensions), coupling with diffusion equations, and links between the level set technique and Brakke's original varifold approach.

On the theoretical/numerical analysis side, level set techniques exploit the considerable technology developed in the area of viscous solutions to

Hamilton–Jacobi equations; see the work in Barles [20] and in Crandall, Evans, Ishii, and Lions [71, 72, 74, 153].

A wide range of applications relate to level set methods, including work on minimal arrival times by Falcone [92], and flame propagation work by Zhu and Ronney [284]. We also refer the reader to a study of various interface techniques in [203] and the collection of papers from the International Conference on Mean Curvature Flow [41]. Fast Marching Methods, while somewhat newer, are also being developed in a host of settings, including applications to CAD/CAM, ruling of surfaces, grid generation, optics, antenna design, and wave dynamics.

This book on Level Set Methods and Fast Marching Methods has aimed to provide some framework for transforming new interface problems into a partial differential equations framework. The field is growing rapidly, and it is with regret that many new developments have been left out. Finally, to repeat something said in the first chapter, this is not the last word on the subject of interfaces. Many other techniques work well in a collection of complex settings. The reader is reminded that, given only a hammer, everything starts to look like a nail. As a general rule, a good, varied, and somewhat dispassionate tool chest is invaluable.

Level Set Methods and Fast Marching Methods require careful thought, as well as a little artistry. The payoff is robust, accurate, and versatile algorithms for highly complex problems.

Acknowledgements

All calculations were performed at the University of California at Berkeley and the Lawrence Berkeley Laboratory. The detailed applications of Level Set and Fast Marching Methods discussed in this work are joint with D. Adalsteinsson, T. Barth, D. Chopp, R. Kimmel, R. Malladi, B. Milne. M. Popovici, C. Rhee, J, Strain, L. Talbot, A. Vladimirsky, J. Wilkening, and J. Zhu.

E-mail may be sent to the author at sethian@math.berkeley.edu. In addition, a web page devoted to interface techniques may be found at

http://math.berkeley.edu/~sethian/level_set.html.

Bibliography

[1] Abgrall, R., *Numerical Discretization of the First-Order Hamilton-Jacobi Equation on Triangular Meshes*, Comm. Pure Appl. Math., 49, pp. 1339–1373, 1996

[2] Adalsteinsson, D., and Sethian, J.A., *A Fast Level Set Method for Propagating Interfaces*, J. Comp. Phys., 118, 2, pp. 269–277, 1995.

[3] Adalsteinsson, D., and Sethian, J.A., *A Unified Level Set Approach to Etching, Deposition and Lithography I: Algorithms and Two-dimensional Simulations*, J. Comp. Phys., 120, 1, pp. 128–144, 1995.

[4] Adalsteinsson, D., and Sethian, J.A., *A Unified Level Set Approach to Etching, Deposition and Lithography II: Three-dimensional Simulations*, J. Comp. Phys., 122, 2, pp. 348–366, 1995.

[5] Adalsteinsson, D., and Sethian, J.A., *A Unified Level Set Approach to Etching, Deposition and Lithography III: Complex Simulations and Multiple Effects*, J. Comp. Phys., 138, 1, pp. 193-223, 1997.

[6] Adalsteinsson, D., and Sethian, J.A., *The Fast Construction of Extension Velocities in Level Set Methods*, 148, 1999, pp. 2-22.

[7] Adalsteinsson, D., Kimmel, R., Malladi, R., and Sethian, J.A., *Fast Marching Methods for Computing the Solutions to Static Hamilton–Jacobi Equations*, CPAM Report 667, Univ. of California, Berkeley, 1996.

[8] Altschuler, S., Angenent, S.B., and Giga, Y., *Mean Curvature Flow through Singularities for Surfaces of Rotation*, J. Geom. Anal.,5,3, pp.293–358, 1995.

[9] Alvarez, L., Lions, P.L., and Morel, M., *Image Selective Smoothing and Edge Detection by Nonlinear Diffusion. II,* SIAM J. Num. Anal. 29, 3, pp. 845–866, 1992.

[10] Alvarez, L., and Mazorra, L., *Signal and Image Restoration using Shock Filters and Anisotropic Diffusion*, SIAM J. Num. Anal., 31, 2, pp. 590–605, 1994.

[11] Ambrosio, L., and Soner, H.M., *Level Set Approach to Mean Curvature Flow in Arbitrary Co-dimension*, J. Diff. Geom, 43, 4, pp. 693–737, 1996.

[12] Aminzadeh, F., Burkhard, N., Long, J., Kunz, T., and Duclos, P., *Three dimensional SEG/EAEG models – an update*, The Leading Edge, 15, pp. 131–136, 1996.

[13] Angenent, S., *Shrinking Doughnuts*, in "Proceedings of Nonlinear

Diffusion Equations and Their Equilibrium States", 3, Eds. N.G. Lloyd et al., Birkhauser, Boston, MA, 1992.

[14] Angenent, S., Ilmanen, T., and Chopp., D.L., T., *A Computed Example of Nonuniqueness of Mean Curvature Flow in R^3* , Comm. Partial Diff. Eqns., 20, 11-1, pp. 1937–1958, 1995.

[15] Arbter, K., Snyder, W.E., Burkhardt, H., and Hirzinger, G., *Application of Affine-invariant Fourier Descriptors to Recognition of 3-D Objects*, IEEE Trans. on Patt. Anal. and Mach. Intell., 12, 7, pp. 640–647, 1990.

[16] Aslam, T., Bzdil, J., and Stewart, D., *Level Set Methods Applied to Modeling Detonation Shock Dynamics*, J. Comp. Phys., 126, pp.390-409, 1996.

[17] Bajcsy, R., and Solina, F., *Three-Dimensional Object Representation Revisited*, in "Proceedings of First International Conference on Computer Vision", pp. 231–240, London, England, 1987.

[18] Bardi, M., and Falcone, M., *An Approximation Scheme for the Minimum Time Function*, SIAM J. Control Optim, 28, pp. 950–965, (1990),

[19] Barles, G., *Remarks on a Flame Propagation Model*, INRIA Report 464, 1985.

[20] Barles, G., *Discontinuous Viscosity Solutions of First Order Hamilton–Jacobi Equations: A Guided Visit*, Non-linear Analysis: Theory, Methods, and Applications, 20, 9, pp. 1123–1134, 1993.

[21] Barles, G., and Georgelin, C., *A simple proof of convergence for an approximation scheme for computing motions by mean curvature*, SIAM J. Numer. Anal., 32, 2, 484–500, 1995.

[22] Barles, G., and Souganidis, P.E., *Convergence of Approximation Schemes for Fully Non-linear Second Order Equations*, Asymptotic Anal., 4, pp. 271–283, 1991.

[23] Barth, T.J., and Sethian, J.A., *Implementation of Hamilton-Jacobi and Level Set Equations on Triangulated Domains*, von Karman Institute Lecture Series, Computational Fluid Mechanics, 1998.

[24] Barth, T.J., and Sethian, J.A., *Numerical Schemes for the Hamilton-Jacobi and Level Set Equations on Triangulated Domains*, J. Comp. Phys., 145, 1, pp. 1–40, 1998.

[25] Bell, J.B., Colella, P., and Glaz, H.M., *A Second-Order Projection Method for the Incompressible Navier-Stokes Equations*, J. Comp. Phys., 85, pp. 257–283, 1989.

[26] Berger, M., and Colella, P., *Local Adaptive Mesh Refinement for Shock Hydrodynamics*, J. Comp. Phys., 1, 82, pp. 62–84, 1989.

[27] Binford, T.O., *Visual Perception by Computer*, invited talk, IEEE Systems and Control Conference, Miami, FL.

[28] Blake, A., and Zisserman, A., *Visual Reconstruction*, MIT Press, Cambridge, MA, 1987.

[29] Blomgren, P. and Chan, T.F., *Color TV: Total Variation Methods for Restoration of Vector-Valued Images*, IEEE Trans. Image Proc., 7,3, pp.304–309, 1998.

[30] Blum, H., *A Transformation for Extracting New Descriptors of Shape*, in "Models for the Perception of Speech and Visual Form", Ed. W. Wathen-Dunn, MIT Press, Cambridge, MA, 1967.

[31] Borgefors, G., *Distance Transformations in Digital Images*, Computer

Vision, Graphics, and Image Processing, 34, pp. 344–371, 1986.

[32] Bourgeois, A., Bourget, M., Lailly, P., Poulet, M., Ricarte, P., and Versteeg R., *Marmousi, model and data*, Proceedings of the 1990 EAEG Workshop on Practical Aspects of Seismic Data Inversion, 1991.

[33] Bourlioux, A., *A Coupled Level-Set Volume of Fluid Algorithm for Tracking Material Interfaces*, Sixth International Symposium on Computational Fluid Dynamics, Sept. 4-8, 1995, Lake Tahoe, NV.

[34] Brackbill, J.U., Kothe, D.B., and Zemach, C., *A Continuum Method for Modeling Surface Tension*, J. Comp. Phys., 100, pp. 335–353, 1992.

[35] Brakke, K.A., *The Motion of a Surface by Its Mean Curvature*, Princeton University Press, Princeton, NJ, 1978.

[36] Brakke , K.A., Surface Evolver Program, Research Report GCC 17, the Geometry Supercomputer Project, University of Minnesota, Minneapolis, MN, 55455, 1990.

[37] Brewer, M.R., *Neural Networks for Meteorological Satellite Image Interpretation*, PhD. Thesis, University of Oxford, 1997.

[38] Brewer, M.R., Malladi, R., Pankiewicz, G., Conway, B., and Tarassenko, L., *Methods for Large-Scale Segmentation of Cloud Images*, 1997 EUMETSAT Meteorological Satellite Data Users' Conference, Brussels, 1997.

[39] Bronsard, L., and Kohn, R.V., *Motion By Mean Curvature as the Singular Limit of Ginzburg-Landau Dynamics*, J. Diff. Eqns., 90, 2, pp. 211–237, 1991.

[40] Bronsard, L., and Wetton, B., *A Numerical Method for Tracking Curve Networks Moving with Curvature Motion*, J. Comp. Phys., 120, 1, pp. 66–87, 1995.

[41] Buttazzo, G., and Visitin, A., *Motion by Mean Curvature and Related Topics*, Proceedings of the International Conference at Trento, 1992, Walter de Gruyter, New York, 1994.

[42] Cahn, J.E., and Hilliard, J.E., *Free energy of a nonuniform system. 1. Interfacial free energy*, Jour. Chem. Phys. 28, pp. 358–367, 1958.

[43] Cale, T.S., Jain, M.K., Tracy, C.J., and Duffin, R., submitted for publication, J. Vac. Sci, Tech, B, 1996,

[44] Cale, T.S., and Raupp, G.B., *Free Molecular Transport and Deposition in Cylindrical Features*, J. Vac. Sci. Tech., B, 8, 4, pp. 649–655, 1990.

[45] Cale, T.S., and Raupp, G.B., *Free Molecular Transport and Deposition in Long Rectangular Trenches*, J. Appl. Phys., 68, 7, pp. 3645–3652, 1990.

[46] Cale, T.S., and Raupp, G.B., *A Unified Line-of-Sight Model of Deposition in Rectangular Trenches*, J. Vac. Sci. Tech., B, 8, 6, pp. 1242–1248, 1990.

[47] Canny, J., *A Computational Approach to Edge Detection*, IEEE Trans. on Patt. Anal. and Mach. Intell., 8, pp. 679–698, 1986.

[48] Carmona, R., *Adaptive Smoothing Respecting Feature Directions*, IEEE Trans. Image Proc., 7,3, pp. 353–358, 1998.

[49] Caselles, V., Catte, F., Coll, T., and Dibos, F., *A Geometric Model for Active Contours in Image Processing*, Numer. Math., 66, pp. 1–31, 1993.

[50] Caselles, V., Kimmel, R., Sapiro, G., *Geodesic active contours*, Proc. Fifth IEEE International Conference on Computer Vision, ICCV '95, pages 694–699, Cambridge, USA, June 1995.

[51] Caselles, V., Morel,J-M, Sbert, C., *An Axiomatic Approach to Image Interpolation*, IEEE Trans. Image Proc., 7,3, pp. 376–386, 1998.

[52]   Castillo, J.E., *Mathematical Aspects of Grid Generation*, Frontiers in Applied Mathematics, 8, SIAM Publications, Philadelphia, PA, 1991.

[53]   Chan, T., and Wong, C-K, *Total Variation Blind Deconvolution*, IEEE Trans. Image Proc., 7,3, pp. 370–375, 1998.

[54]   Chang, Y.C., Hou, T.Y., Merriman, B., and Osher, S.J., *A Level Set Formulation of Eulerian Interface Capturing Methods for Incompressible Fluid Flows*, Jour. Comp. Phys., 124, pp. 449-464, 1996.

[55]   Chen, S., Merriman, B., Osher, S., and Smereka, P., *A Simple Level Set Method for Solving Stefan Problems*, J. Comp. Phys., 138, pp. 8–29, 1997.

[56]   Chen, Y., Bi, Y., and Jiang, T., *The Liquid Bridge with Marangoni Effect*, Communications in Nonlinear Science and Numerical Simulation, 1, 1, pp. 48–51, 1996.

[57]   Chen, Y., Giga, Y., and Goto, S., *Uniqueness and Existence of Viscosity Solutions of Generalized Mean Curvature Flow Equations*, J. Diff. Geom, 33, 749, 1991.

[58]   Chew, L.P., *Guaranteed-Quality Triangular Meshes*, Technical Report, TR 89-983, Cornell University Department of Computer Science, March, 1989

[59]   Chopp, D.L., *Computing Minimal Surfaces via Level Set Curvature Flow*, Jour. of Comp. Phys., 106, pp. 77–91, 1993.

[60]   Chopp, D.L., *Numerical Computation of Self-Similar Solutions for Mean Curvature Flow* J. Exper. Math., 3, 1, pp. 1–15, 1994.

[61]   Chopp, D.L., and Sethian, J.A., *Flow Under Curvature: Singularity Formation, Minimal Surfaces, and Geodesics*, Jour. Exper. Math., 2, 4, pp. 235–255, 1993.

[62]   Chopp, D.L., and Sethian, J.A., *Motion by Intrinsic Laplacian of Curvature*, CPAM Report PAM-746, Dept. of Mathematics, Univ. of California, Berkeley,Sept. 1998, accepted for publication, Nov. 1998, in press, Interfaces and Free Boundaries, 1999.

[63]   Chorin, A.J., *Numerical Solution of the Navier-Stokes Equations*, Math. Comp., 22, pp. 745, 1968.

[64]   Chorin, A.J., *Numerical Study of Slightly Viscous Flow*, J. Fluid Mech., 57, pp. 785–796, 1973.

[65]   Chorin, A.J., *Flame Advection and Propagation Algorithms*, J. Comp. Phys., 35, pp. 1–11, 1980.

[66]   Chorin, A.J., *Curvature and Solidification*, J. Comp. Phys., 57, pp. 472–490, 1985.

[67]   Chorin, A.J., and Marsden, J.E., *A Mathematical Introduction to Fluid Mechanics*, Springer-Verlag, New York, NY, 1980.

[68]   Cohen, L.D., *On Active Contour Models and Balloons*, Computer Vision, Graphics, and Image Processing, 53, 2, pp. 211–218, 1991.

[69]   Colella, P., and Puckett, E.G., *Modern Numerical Methods for Fluid Flow*, Lecture Notes, Department of Mechanical Engineering, University of California, Berkeley, CA, 1994.

[70]   Cottet, G.H., and El Ayyadi, M., *A Volterra Type Model for Image Processing*, IEEE Trans. Image Proc., 7,3, pp. 292–303, 1998.

[71]   Crandall, M.G., Evans, L.C., and Lions, P-L., *Some Properties of Viscosity Solutions of Hamilton–Jacobi Equations*, Tran. AMS, 282, pp. 487–502, 1984.

[72]   Crandall, M.G., Ishii, H., and Lions, P-L., *User's Guide to Viscosity Solutions of Second Order Partial Differential Equations*, Bull. AMS, 27/1,

pp. 1–67, 1992.

[73]   Crandall, M.G. and Lions, P.L., Two Approximations of Solutions of Hamilton-Jacobi Equations, Math. Comp., 167, 43, pp. 1–19, 1984

[74]   Crandall, M.G., and Lions, P-L., *Viscosity Solutions of Hamilton–Jacobi Equations*, Tran. AMS, 277, pp. 1–43, 1983.

[75]   Crimmins, T., *A Complete Set of Fourier Descriptors for Two-dimensional Shapes*, IEEE Trans. on Syst. Man, and Cyber., 12, 6, 1982.

[76]   Danielson, P.E., *Euclidean distance mapping*, Computer Graphics and Image Processing, 14, pp. 227–248, 1980.

[77]   Deconinck, H., Struijs, R., and Roe, P.L., *Compact Advection Schemes on Unstructured Grids*, von Karman Institute Lecture Series 1988-05, VKI LS 1993-04, Computational Fluid Dynamics, 1993

[78]   Dijkstra, E.W., *A Note on Two Problems in Connection with Graphs*, Numerische Mathematic, 1:269–271, 1959.

[79]   Dziuk, Gerhard, *An Algorithm for Evolutionary Surfaces*, Num. Math. 58, pp. 603–611, 1991.

[80]   Ecker, K., Huisken, G., *Interior Estimates for Hypersurfaces Moving by Mean Curvature*, Inventiones Mathematica, 105, 3, pp. 547–569, 1991.

[81]   Eiseman, P.R., *Grid Generation for Fluid Mechanics Computations*, Ann. Rev. Fluid Mech., 17, pp. 487–522, 1985.

[82]   Elliot, D.F. and Rao, K.R., *Fast Transforms: Algorithms, Analyses, Applications*, Academic Press, New York, NY, 1982.

[83]   Engquist, B., and Osher, S.J., *Stable and Entropy-Satisfying Approximations for Transonic Flow Calculations*, Math. Comp., 34, 45, 1980.

[84]   Escher, J., Mayer, U., and Simonett, G., *On the Surface Diffusion Flow*, Proc. Intern. Conf. on Navier Stokes Equations and Related Problems, TEV/VSP, Vilnius/Utrecht, 1998.

[85]   Evans, L.C., *Convergence of an algorithm for mean curvature motion*, Indiana Univ. Math. J., 42, 553-557, 1993.

[86]   Evans, L.C., *Partial Differential Equations*, Berkeley Mathematics Lecture Notes Series, Vol. 3A, 3B, Center for Pure and Applied Mathematics, University of California, Berkeley, CA, 1994.

[87]   Evans, L.C., Soner, H.M., and Souganidis, P.E., *Phase Transitions and Generalized Motion by Mean Curvature*, Communications on Pure and Applied Mathematics, 45, pp. 1097–1123, 1992.

[88]   Evans, L.C., and Spruck, J., *Motion of Level Sets by Mean Curvature I*, J. Diff. Geom, 33, 635, 1991.

[89]   Evans, L.C., and Spruck, J., *Motion of Level Sets by Mean Curvature II*, Transactions of the American Mathematical Society, 330, 1, pp. 321–332, 1992.

[90]   Evans, L.C., and Spruck, J., *Motion of Level Sets by Mean Curvature III*, J. Geom. Anal. 2, pp. 121–150, 1992.

[91]   Evans, L.C., and Spruck, J., *Motion of Level Sets by Mean Curvature IV*, J. Geom. Anal., 5, 1, pp. 77–114, 1995.

[92]   Falcone, M., *The Minimum Time Problem and Its Applications to Front Propagation*, in "Motion by Mean Curvature and Related Topics", Proceedings of the International Conference at Trento, 1992, Walter de Gruyter, New York, 1994.

[93]   Falcone, M., Giorgi, T., and Loretti, P., *Level Sets of Viscosity*

Solutions: Some Applications to Fronts and Rendez-Vous Problems,
SIAMJ. Appl. Math., 54, 5, pp. 1335–1354, 1994.

[94]   Fatemi, E., Engquist, B., and Osher, S.J., *Numerical Solution of the High Frequency Asymptotic Wave Equation for the Scalar Wave Equation,* J. Comp. Phys,. 120, pp. 145–155, 1995.

[95]   Faugeras, O., and Keriven, R., *Variational Principles, Surface Evolution, PDE's, Level Set Methods, and the Stereo Problem,* IEEE Trans. Image Proc., 7, 3, pp. 336–344, 1998.

[96]   H. Freeman, *On the Encoding of Arbitrary Geometric Configurations,* IEEE Trans. on Electronic Computers, EC-10, pp. 260–268, 1961.

[97]   Gage, M., *Curve Shortening Makes Convex Curves Circular,* Inventiones Mathematica, 76, pp. 357, 1984.

[98]   Gage, M., and Hamilton, R., *The Equation Shrinking Convex Planes Curves,* J. Diff. Geom, 23, pp. 69, 1986.

[99]   Garabedian, P., Partial Differential Equations, Wiley, New York, 1964.

[100]   Ghoniem, A.F., Chorin, A.J., and Oppenheim, A.K., *Numerical Modeling of Turbulent Flow in a Combustion Tunnel,* Philos. Trans. Roy. Soc. Lond. A., 304, pp. 303–325, 1982.

[101]   Giga, Y., and Goto, S., *Motion of Hypersurfaces and Geometric Equations,* Journal of the Mathematical Society of Japan, 44, pp. 99, 1992.

[102]   Giga, Y., Goto, S., Ishii, H., *Global Existence of Weak Solutions for Interface Equations Coupled with Diffusion Equations,* SIAM J. Math. Anal., 23, N4, pp. 821–835, 1992.

[103]   Girao P.M., Kohn, R.V., *Convergence of a Crystalline Algorithm for the Heat Equation in One Dimension and for the Motion of a Graph by Weighted Curvature,* Num. Math., 67, 1, pp. 41–70. 1994.

[104]   Gonzalez, R. C., and Wintz, P., *Digital Image Processing (2nd Ed.),* Addison-Wesley, Reading, MA, 1987.

[105]   Grayson, M., *The Heat Equation Shrinks Embedded Plane Curves to Round Points,* J. Diff. Geom., 26, pp. 285, 1987.

[106]   Grayson, M., *A Short Note on the Evolution of Surfaces Via Mean Curvatures,* J. Diff. Geom., 58, pp. 555, 1989.

[107]   Greengard, L., and Strain, J., *A Fast Algorithm for Evaluating Heat Potentials,* Comm. Pure Appl. Math., XLIII, pp. 949–963, 1990.

[108]   Guichard, F., *A Morphological, Affine, and Galilean Invariant Scale-Space for Movies,* IEEE Trans. Image Proc., 7,3, pp. 444-456, 1998.

[109]   Gurtin, M.E., *On the Two-Phase Stefan Problem with Interfacial, Energy and Entropy,* Arch. Rat. Mech. Anal., 96, pp. 199–241, 1986.

[110]   Harten, A., Engquist, B., Osher, S., and Chakravarthy, S., *Uniformly High Order Accurate Essentially Non-oscillatory Schemes. III,* J. Comp. Phys., 71, 2, pp. 231–303, 1987.

[111]   Hayes, W.D., *The Vorticity Jump Across a Gasdynamic Discontinuity,* J. Fluid Mech., 2, pp. 595–600, 1959.

[112]   Helmsen, J.J., *A Comparison of Three-Dimensional Photolithography Development Methods,* Ph.D. Dissertation, EECS, University of California, Berkeley, CA, 1994.

[113]   Hirt, C.W., and Nicholls, B.D., *Volume of Fluid (VOF) Method for Dynamics of Free Boundaries,* J. Comp. Phys., 39, pp. 201–225, 1981.

[114]   Holm, E., and Langtangen, H., *A Method for Simulating Sharp Fluid Interfaces in Groundwater Flow,* submitted for publication, Advances in Water Resouces, April, 1998.

[115] Horn, B.K.P., and Brooks, Eds., Shape from Shading, MIT Press, Cambridge, MA, 1989.

[116] Hou, T.Y., Li, Zhilin, L., Osher, S.J., and Zhao. H.K., *A Hybrid Method for Moving Interfaces Problems with Application to the Hele-Shaw Flow*, J. Comp. Phys., 134, 2, pp. 236–52, 1997.

[117] Huisken, G., *Flow by Mean Curvature of Convex Surfaces into Spheres*, J. Diff. Geom., 20, pp. 237, 1984.

[118] Huisken, G., *Asymptotic Behavior for Singularities of the Mean Curvature Flow*, J. Diff. Geom., 31, pp. 285–299, 1991.

[119] Ilmanen, T., *Generalized Flow of Sets by Mean Curvature on a Manifold*, Indiana University Mathematics Journal, 41, 3, pp. 671–705, 1992.

[120] Ilmanen, T., *Elliptic Regularization and Partial Regularity for Motion by Mean Curvature*, Memoirs of the American Mathematical Society, 108, 520, 1994.

[121] Jain, M.K., Cale, T.S., Tracy, C.J., and Duffin, R.L., *Curvature Driven Surface Diffusion of Aluminum-(1.5)Copper During Sputter Deposition*, Proceedings, 11th VMIC Conference, 1992.

[122] Johnson, C., *Numerical Solution of Partial Differential Equations by the Finite Element Method*, Cambridge University Press, Cambridge, 1987.

[123] Kass, M., Witkin, A., and Terzopoulos, D., *Snakes: Active Contour Models*, International Journal of Computer Vision, pp. 321–331, 1988.

[124] Katardjiev, I.V., Carter, G., Nobes, M.J., *Precision Modeling of the Mask-Substrate Evolution During Ion Etching*, J. Vac. Science Technology, A 6, 4, pp. 2443–2450, 1988.

[125] Kelly, F.X., and Ungar, L.H., *Steady and Oscillatory Cellular Morphologies in Rapid Solidification*, Phys. Rev. B, 34, pp. 1746–1753, 1986.

[126] Kessler, D.A., and Levine, H., *Stability of Dendritic Crystals*, Phys. Rev. Lett., 57, pp. 3069–3072, 1986.

[127] Kichenassamy, S., Kumar, A., Olver, P., Tannenbaum, A., and Yezzi, A., *Gradient Flows and Geometric Active Contours*, Proc. Fifth IEEE International Conference on Computer Vision, ICCV '95, pages 810–815, Cambridge, USA, June 1995.

[128] Kimmel, R., *Curve Evolution on Surfaces*, Ph.D. Thesis, Dept. of Electrical Engineering, Technion, Israel, 1995.

[129] Kimmel, R., Amir, A., and Bruckstein, A.M., *Finding Shortest Paths on Surfaces Using Level Sets Propagation*, IEEE Trans. Patt. Anal. Machine Intell., 17, 6, pp. 635–640, 1995.

[130] Kimmel, R., and Bruckstein, A., *Shape from Shading via Level Sets*, Center for Intelligent Systems Report No. 9209, Technion-Israel Institute of Technology, June 1992.

[131] Kimmel, R., and Bruckstein, A., *Shape Offsets via Level Sets*, Computer-Aided Design, 25, 3, pp. 154–161, 1993.

[132] Kimmel, R., Malladi, R., and Sochen, N., *Images as Embedded Maps and Minimal Surfaces: Movies, Color, Texture, and Medical Images*, to appear, International Journal of Computer Vision, 1999.

[133] Kimmel, R., and Sapiro, G., *Shortening Three-Dimensional Curves via Two-Dimensional Flows*, Compter. Math. Applic, 29, 3, pp. 49–62, 1995.

[134] Kimmel, R., and Sethian, J.A., *Fast Marching Methods for*

Computing Distance Maps and Shortest Paths, CPAM Report 669, Univ. of California, Berkeley, 1996.

[135] Kimmel, R., and Sethian, J.A., *Fast Marching Methods for Robotic Navigation with Constraints*, Center for Pure and Applied Mathematics Report, Univ. of California, Berkeley, May 1996, submitted for publication, Int. Journal Robotics Research, 1998.

[136] Kimmel, R., and Sethian, J.A., *An Optimal Time Algorithm for Shape from Shading*, LBNL-41660, Lawrence Berkeley National Laboratory, Berkeley, California, April 1998.

[137] Kimmel, R., and Sethian, J.A., *Fast Marching Methods on Triangulated Domains*, Proc. Nat. Acad. Sci., 95, pp. 8341-8435, 1998.

[138] Kimmel, R., and Sethian, J.A., *Fast Voronoi Diagrams and Offsets on Triangulated Surfaces*, submitted for publication, CAD Special issue on Offsets, Sweeps, and Minkowsky Sums, G. Elber, Ed., July, 1998.

[139] Kimmel, R., Shaked, D., Kiryati, N., and Bruckstein, A.M., *Skeletonization via Distance Maps and Level Sets*, Computer Vision and Image Understanding (CVIU) 62, 4, pp. 382–391, 1995.

[140] Knupp, P.M., and Steinberg, S., *The Fundamentals of Grid Generation*, preprint 1993.

[141] Kuiken, H.K., *Viscous Sintering: the Surface-tension-driven Flow of a Liquid Form under the Influence of Curvature Gradients at its Surface*, J. Fluid Mech., 214, pp. 503–515, 1990.

[142] Lafaurie, B., Nardone, C., Scardovelli, R., Zaleski, S., and Zanetti, G., *Modelling Merging and Fragmentation in Multiphase Flows with SURFER*, J. Comp. Phys, 113, 1, pp. 134–47, 1994.

[143] Langer, J.S., *Instabilities and Pattern Formation in Crystal Growth*, Rev. Mod. Phys., 52, pp. 1–28, 1980.

[144] Latombe, J.C., *Robot motion planning*, Kluwer Academic Publishers, Boston, MA, 1991.

[145] Lax, P.D., *Hyperbolic Systems of Conservation Laws and the Mathematical Theory of Shock Waves*, SIAM Reg. Conf. Series, Lectures in Applied Math, 11, pp. 1–47, 1970.

[146] Lee, D.T., *Medial Axis Transformation of a Planar Shape*, IEEE Trans. Patt. Anal. Machine Intell., 4, pp. 363–369, 1982.

[147] Leon, F.A., Tazawa, S., Saito, K., Yoshi, A., and Scharfetter, D.L., *Numerical Algorithms for Precise Calculation of Surface Movement in 3-D Topography Simulation*, 1993 International Workshop on VLSI Process and Device Modeling: VPAD.

[148] Litman, A., Lesselier, D., and Santosa, F., *Reconstruction of a Two-dimensional Binary Obstacle by Controlled Evolution of a Level set*, Inverse Problems, 13, pp. 685–706, 1998.

[149] LeVeque, R.J., *Numerical Methods for Conservation Laws*, Birkhauser, Basel, 1992.

[150] LeVeque, R.J., and Li, Z., *The Immersed Interface Method for Elliptic Equations with Discontinuous Coefficients and Singular Sources*, SIAM J. Num. Anal., 13, pp. 1019–1044, 1994.

[151] Leymarie, F., and Levine, M.D., *Simulating the Grassfire Transform using an Active Contour Model*, IEEE Trans. Patt. Anal. Machine Intell., 14, 1, pp. 56–75, 1992.

[152] Li, X.L., *Study of Three-Dimensional Rayleigh-Taylor Instability in Compressible Fluids Through Level Set Method and Parallel Computation*,

Phys. Fluids A, 5, 1, pp. 1904–1913, 1993.

[153]   Lions, P.L., *Generalized Solution of Hamilton–Jacobi Equations*, Pittman, London, 1982.

[154]   Löhner, R.L. and Baum, J.D., *Numerical Simulation of Shock Interaction with Complex Geometry Structures Using a New h-Refinement Scheme on Unstructured Grids*, 28th AIAA Aerospace Sciences Mtg, 1990

[155]   Lorenson, W.E., and Cline, H.E., *Marching Cubes: A High Resolution 3D Surface Construction Algorithm*, Computer Graphics, 21, 4, 1987.

[156]   Majda, A., and Sethian, J.A., *Derivation and Numerical Solution of the Equations of Low Mach Number Combustion*, Combustion Science and Technology, 42, pp. 185–205, 1984.

[157]   Malladi, R., *Constrained Flows*, LBL Internal Report, Lawrence Berkeley National Laboratory, University of California, 1997.

[158]   Malladi, R., and Sethian, J.A., *A Unified Approach for Shape Segmentation, Representation, and Recognition*, Center for Pure and Applied Mathematics, Report 614, University of California, Berkeley, 1994

[159]   Malladi, R., and Sethian, J.A., *Image Processing via Level Set Curvature Flow*, Proc. Natl. Acad. of Sci., 92, 15, pp. 7046–7050, 1995.

[160]   Malladi, R., and Sethian, J.A., *Level Set Methods for Curvature Flow, Image Enhancement, and Shape Recovery in Medical Images*, Proc. of Conf. on Visualization and Mathematics, June, 1995, Berlin, Germany, Springer-Verlag, Heidelberg, Germany, 1997.

[161]   Malladi, R., and Sethian, J.A., *Image Processing: Flows under Min/Max Curvature and Mean Curvature*, Graphical Models and Image Processing, 58,2, pp. 127–141, 1996.

[162]   Malladi, R., and Sethian, J.A., *A Unified Approach to Noise Removal, Image Enhancement, and Shape Recovery*, IEEE Trans. on Image Processing, 5, 11, pp. 1554–68, 1996.

[163]   Malladi, R., Sethian, J.A., and Vemuri, B.C., *A Fast Level Set based Algorithm for Topology-Independent Shape Modeling* J. Math. Imaging and Vision, 6, 2/3, pp. 269–290, 1996.

[164]   Malladi, R., and Sethian, J.A., *An $O(N \log N)$ Algorithm for Shape Modeling*, Proc. Nat. Acad. Sci., Vol. 93, pp. 9389-9392, 1996.

[165]   Malladi, R., and Sethian, J.A., *Shape Modeling in Medical Imaging with Marching Methods*, Report LBNL-39541, LBNL, University of California, Berkeley, Oct. 1996.

[166]   Malladi, R., and Sethian, J.A., *Level Set and Fast Marching Methods in Image Processing and Computer Vision*, Proceedings of IEEE International Conference on Image Processing, Lausanne, Switzerland, Sept. 16-19, 1996.

[167]   Malladi, R., and Sethian, J.A., *An $O(NlogN)$ Algorithm for Shape Modeling*, with R. Malladi, Proceedings of the National Academy of Sciences, Vol. 93, pp. 9389-9392, September 1996.

[168]   Malladi, R., Sethian, J.A., and Vemuri, B.C., *Evolutionary Fronts for Topology-independent Shape Modeling and Recovery*, in Proceedings of Third European Conference on Computer Vision, Stockholm, Sweden, Lecture Notes in Computer Science, 800, pp. 3–13, 1994.

[169]   Malladi, R., Sethian, J.A., and Vemuri, B.C., *Shape Modeling with Front Propagation: A Level Set Approach*, IEEE Trans. on Pattern Analysis and Machine Intelligence, 17, 2, pp. 158–175, 1995.

[170]   Markstein, G.H., *Nonsteady Flame Propagation*, Pergamon Press,

1964.

[171]   Marr, D., and Hildreth, E., *A Theory of Edge Detection*, Proc. of Royal Soc. (London), B207, pp. 187–217, 1980.

[172]   Mayya, N., and Rajan, V.T., "Voronoi Diagrams of Polygons: A Framework for Shape Representation,", Jour. Math. Imaging and Vision, 6, 4, pp. 355–378, 1996.

[173]   McVittie, J.P., Rey, J.C., Bariya, A.J., et al., *SPEEDIE: A Profile Simulator for Etching and Deposition*, Proceedings of the SPIE - The International Society for Optical Engineering, 1392, pp. 126–38, 1991.

[174]   McVittie, J.P., Rey, J.C., Cheng, L.Y., and IslamRaja, M.M, *LPCVD Profile Simulation Using a Re-emission Model*, IEEE International Electron Devices Meeting 1990. Technical Digest, New York, NY, pp. 917–20, 1990.

[175]   Meiron, D.I., *Boundary Integral Formulation Of The Two-Dimensional, Symmetric Model Of Dendritic Growth*, Physica D, 23, pp. 329–339, 1986.

[176]   Merriman, B., Bence, J., and Osher, S.J., *Motion of Multiple Junctions: A Level Set Approach*, Jour. Comp. Phys., 112, 2, pp. 334–363, 1994.

[177]   Meyer, G. H., *Multidimensional Stefan Problems*, SIAM J. Num. Anal., 10, pp. 552–538, 1973.

[178]   Milne, B. *Adaptive Level Set Methods Interfaces*, PhD. Thesis, Dept. of Mathematics, University of California, Berkeley, CA., 1995.

[179]   Moisan, L., *Affine Plane Curve Evolution: A Fully Consistent Scheme*, IEEE Trans. Image Proc., 7, 3, pp. 411-420, 1998.

[180]   Mori, S., Suen, C.Y., and Yamamoto, K., *Historical review of OCR research and development*, Proc. of the IEEE, 80, 7, pp. 1029–1057, 1992.

[181]   Mulder, W., Osher, S.J., Sethian, J.A., *Computing Interface Motion in Compressible Gas Dynamics*, Jour. Comp. Phys., 100, pp. 209–228, 1992.

[182]   Mullins, W.W., and Sekerka, R.F., *Morphological Stability of a Particle Growing by Diffusion or Heat Flow*, Jour. Appl. Phys., 34, pp.323–329, 1963

[183]   Noh, W., and Woodward, P., *A Simple Line Interface Calculation.* Proceedings, Fifth International Conference on Fluid Dynamics, Eds. A.I. vn de Vooran and P.J. Zandberger, Springer-Verlag, 1976.

[184]   Ogniewicz, R., and Ilg, M., *Voronoi Skeletons: Theory and Application*, Proc. of Conf. on Comp. Vis. and Patt. Recog., Champaign, IL, pp. 63–69, 1992.

[185]   Oliker, V., *Evolution of Non-parametric Surfaces with Speed Depending on Curvature: I, The Gauss Curvature Case*, Indiana Univ. Math. Journal, 40, 1, pp. 237–258, 1991.

[186]   Osher, S., and Rudin, L.I., *Feature-oriented Image Enhancement Using Shock Filters*, SIAM J. Num. Anal., 27, pp. 919–940, 1990.

[187]   Osher, S., and Sethian, J.A., *Fronts Propagating with Curvature-Dependent Speed: Algorithms Based on Hamilton–Jacobi Formulations*, Journal of Computational Physics, 79, pp. 12–49, 1988.

[188]   Osher, S., and Shu, C., *High-Order Nonoscillatory Schemes for Hamilton–Jacobi Equations*, Jour. Comp. Phys., 28, pp. 907–922, 1991.

[189]   Paragios, N., and Deriche, R., *A PDE-based Level-Set Approach for Detection and Tracking of Moving Objects*, INRIA preprint 3173, Institut

National de Recherche en Informatique et en Automatic, May, 1997.

[190] Pasch, E., *The level set method for mean curvature flow on* (R^3, g), Preprint, SFB 382, Mathematisches Insitut, Universitat Tuebingen,

[191] Pavlidis, T., *Polygonal Approximations by Newton's Method*, IEEE Trans. on Computers, C-26, 8, pp. 800–807, 1977.

[192] Pearson, E., and Fu, K.S., *Shape Discrimination Using Fourier Descriptors*, IEEE Trans. System, Man, and Cyber., SMC-7, 3, pp. 170–179, 1977.

[193] Perona, P., *Orientation Diffusions*, IEEE Trans. Image Proc., 7,3, pp. 457–467, 1998.

[194] Perona, P., and Malik, J., *Scale-space and Edge Detection Using Anisotropic Diffusion*, IEEE Trans. Pattern Analysis and Machine Intelligence, 12, 7, pp. 629–639, 1990.

[195] Pimienta, P.J.P., Garboczi, E. J., and Carter, W. C., *Cellular Automaton Algorithm for Surface Mass Transport due to Curvature Gradients: Simulations of Sintering* Comp. Materials Science, 1, pp. 63–77, 1992.

[196] Pindera, M.Z., and Talbot, L., *Flame-Induced Vorticity: The Effects of Stretch*, Twenty-First Symposium (Int'l) on Combustion, The Combustion Institute, Pittsburgh, PA, pp. 1357–1366, 1986.

[197] Press, W.H., Teukolsky, S.A., Vetterling, W.T., and Flannery, B.P., *Numerical Recipes*, Cambridge University Press, New York, 1988.

[198] Puckett, E.G., *A Volume-of-Fluid Interface Tracking Algorithm with Applications to Computing Shock Wave Refraction*, Proceedings of the 4th International Symposium on Computational Computational Fluid Dynamics, Davis, California, 1991.

[199] Reitich, F., and Soner, H.M., *Three-phase boundary motions under constant velocities. I. The vanishing surface tension limit.*, Proc. Royal. Soc. Edin., A., 126, 4, 837-865, 1996.

[200] Rietveld, W. E. A. and Berkhout, A. J., *Prestack depth migration by means of controlled illumination*, Geophysics, 59, 5, pp. 801–809, 1994.

[201] Rey, J.C., Lie-Yea Cheng, McVittie, J.P., and Saraswat, K.C., *Monte Carlo Low Pressure Deposition Profile Simulations*, Journal of Vacuum Science and Technology A (Vacuum, Surfaces, and Films), May-June 1991, 9, 3, 1, pp. 1083–1087.

[202] Rhee, C., Talbot, L., and Sethian, J.A., *Dynamical Study of a Premixed V flame*, Jour. Fluid Mech., 300, pp. 87–115, 1995.

[203] Rider, W.J., and Kothe, D.B., *Stretching and Tearing Interface Tracking Methods*, 12th AIAA CFD Conference, AIAA-95-1717, San Diego, CA., June 20, 1995

[204] Roe, P.L., *Linear Advection Schemes on Triangular Meshes*, CoA 8720, Cranfield Institute of Technology, 1987

[205] Roe, P.L., *"Optimum" Upwind Advection on a Triangular Mesh*, ICASE 90-75, 1990

[206] ter Haar Romeny, B., Ed., Geometry Driven Diffusion in Computer Vision, Kluwer, 1994.

[207] Rouy, E. and Tourin, A., *A Viscosity Solutions Approach to Shape-From-Shading*, SIAM J. Num. Anal, 29, 3, pp. 867–884, 1992.

[208] Rudin, L., Osher, S., and Fatemi, E., *Nonlinear Total Variation-Based Noise Removal Algorithms*, Modelisations Matematiques pour le traitement d'images, INRIA, pp. 149–179, 1992.

[209] Ruppert, J., *A New and Simple Algorithm for Quality Two-Dimensional Mesh Generation*, UCB/CSD 92/694, University of California, Berkeley, Dept. of Computer Science, 1992

[210] W. B. Ruskai et al., *Wavelets and their Applications*, Jones and Barlett Publishers, Boston, MA 1992.

[211] Ruuth, S.J., *Efficient Algorithms for Diffusion-Generated Motion by Mean Curvature*, J. Comp. Phys., pp. 144, 2, pp. 603–625, 1998.

[212] Ruuth, S.J., and Merriman, B., *Convolution Generated Motion and Generalized Huyghens' Principles for Interface Motion*, preprint, Dept. of Mathematics, UCLA, 1998.

[213] Santosa, F., *A Level Set Approach for Inverse Problems Involving Obstacles*, ESIAM Control Optimization and Calculus of Variations, 1, pp. 17–33, 1996.

[214] Sapiro, G., and Tannenbaum, A., *Affine Invariant Scale-Space*, Int. Jour. Comp. Vision, 11, 1, pp. 25–44, 1993.

[215] Sapiro, G., and Tannenbaum, A., *Image Smoothing Based on Affine Invariant Flow*, Proc. of the Conference on Information Sciences and Systems, Johns Hopkins University, March 1993.

[216] Sarti, A., Ortiz, C., Lockett, S., and Malladi, R., *A Unified Geometric Model for 3D Confocal Image Analysis in Cytology*, LBL Report, Lawrence Berkeley National Laboratory, University of California, May, 1998, SIBGRAPI 98 proceedings, Rio de Janeiro, 1998, submitted to IEEE Trans. on Biomedical Engineering.

[217] Scheckler, E.W., Ph.D. Dissertation, EECS, University of California, Berkeley, CA, 1991.

[218] Scheckler, E.W., Toh, K.K.H., Hoffstetter, D.M., and Neureuther, A.R., *3D Lithography, Etching and Deposition Simulation*, Symposium on VLSI Technology, Oiso, Japan, pp. 97–98, 1991.

[219] Schmidt, Alfred, *Computation of Three-dimensional Dendrites with Finite Elements*, 125, pp. 293–312, 1996.

[220] Schneider, W.A. Jr., *Robust and efficient upwind finite-difference traveltime calculations in three dimensions*, Geophysics, 60, pp. 1108–1117, 1995.

[221] Sedgewick, R., *Algorithms*, Addison-Wesley, Reading, MA, 1988.

[222] Sethian, J.A., *An Analysis of Flame Propagation*, Ph.D. Dissertation, Dept. of Mathematics, University of California, Berkeley, CA, 1982.

[223] Sethian, J.A., *The Wrinkling of a Flame Due to Viscosity*, Fire Dynamics and Heat Transfer, Eds. J.G. Quintiere, R.A. Alpert and R.A. Altenkirch, HTD, ASME, New York, NY, 25, pp. 29–32, 1983.

[224] Sethian, J.A., *Turbulent Combustion in Open and Closed Vessels*, J. Comp. Phys., 54, pp. 425–456, 1984.

[225] Sethian, J.A., *Curvature and the Evolution of Fronts*, Comm. in Math. Phys., 101, pp. 487–499, 1985.

[226] Sethian, J.A., *Numerical Methods for Propagating Fronts*, in Variational Methods for Free Surface Interfaces, Eds. P. Concus and R. Finn, Springer-Verlag, NY, 1987.

[227] Sethian, J.A., *Parallel Level Set Methods for Propagating Interfaces on the Connection Machine*, Unpublished manuscript, 1989.

[228] Sethian, J.A., *Numerical Algorithms for Propagating Interfaces: Hamilton–Jacobi Equations and Conservation Laws*, Journal of Differential Geometry, 31, pp. 131–161, 1990.

[229] Sethian, J.A., *A Brief Overview of Vortex Methods*, in Vortex Methods and Vortex Motion, Eds. K. Gustafson and J.A. Sethian, SIAM Publications, Philadelphia, PA, 1991.

[230] Sethian, J.A., *Curvature Flow and Entropy Conditions Applied to Grid Generation*, J. Comp. Phys., 115, pp. 440–454, 1994.

[231] Sethian, J.A., *Algorithms for Tracking Interfaces in CFD and Material Science*, Annual Review of Computational Fluid Mechanics, 1995.

[232] Sethian, J.A., *Level Set Techniques for Tracking Interfaces: Fast Algorithms, Multiple Regions, Grid Generation and Shape/Character Recognition*, Proceedings of the International Conference on Curvature Flows and Related Topics, Trento, Italy, 1994, Eds. A. Damlamian, J. Spruck, and A. Visintin, Gakuto Intern. Series, Tokyo, Japan, 5, pp. 215–231, 1995.

[233] Sethian, J.A., *A Fast Marching Level Set Method for Monotonically Advancing Fronts*, Proc. Nat. Acad. Sci., 93, 4, pp.1591–1595, 1996.

[234] Sethian, J.A., *Fast Marching Level Set Methods for Three-Dimensional Photolithography Development*, Proceedings, SPIE 1996 International Symposium on Microlithography, Santa Clara, California, March, 1996.

[235] Sethian, J.A., *A Review of the Theory, Algorithms, and Applications of Level Set Methods for Propagating Interfaces*, Acta Numerica, Cambridge University Press, 1996.

[236] Sethian, J.A., Level Set Methods: Evolving Interfaces in Geometry, Fluid Mechanics, Computer Vision, and Materials Sciences, First Edition, Cambridge University Press, 1996.

[237] Sethian, J.A., *Tracking Interfaces with Level Sets*, American Scientist, pp. 254–263, May–June, 1997.

[238] Sethian, J.A., *Fast Marching Methods and Level Set Methods for Propagating Interfaces* von Karman Institute Lecture Series, Computational Fluid Mechanics, 1998.

[239] Sethian, J.A., *Fast Marching Methods*, SIAM Review, 41, July, 1999.

[240] Sethian, J.A., *Algorithms for Multi-Valued Solutions of the Eikonal Equation*, in progress, 1999.

[241] Sethian, J.A., and Adalsteinsson, D., *An Overview of Level Set Methods for Etching, Deposition, and Lithography Development*, IEEE Transactions on Semiconductor Devices, 1996. 10, 1, pp.167-184, 1997.

[242] Sethian, J.A., and Popovici, M., *Fast Marching Methods Applied to Computation of Seismic Travel Times*, Geophysics, 64, 2, 1999.

[243] Sethian, J.A. and Strain, J.D., *Crystal Growth and Dendritic Solidification* J. Comp. Phys., 98, pp. 231–253, 1992.

[244] Sethian, J.A., and Vladimirsky, A., *Extensions to Triangulated Fast Marching Methods*, to be submitted for publication, 1999.

[245] Sethian, J.A., and Wilkening, J., *Interface Tracking Techniques for Electromigration and Metallization Failure*, to be submitted for publication, 1998.

[246] Sherwin, W., Karniadakis, G.E., and Orszag, S.A,. *Numerical Simulation of the Ion Etching Process*, J. Comp. Phys., 110, 2, pp. 373–398, 1994.

[247] Smiljanovksi, V., Moser, V., and Klein, R., *A Capturing-Tracking Hybrid Scheme for Deflagration Discontinuities*, Combustion Theory and Modelling, 1, 183-216, 1997

[248] Singh, V.K., Shaqfeh, S.G., and McVittie, J.P., *Simulation of Profile Evolution in Silicon Reactive Ion Etching with Re-emission and Surface Diffusion*, J. Vac. Sci. Tech., B. 10, 3, pp. 1091–1104, 1993.

[249] Smith, J. B., *Shape Instabilities and Pattern Formation in Solidification: A New Method for Numerical Solution of the Moving Boundary Problem*, Jour. Comp. Phys., 39, pp. 112–127, 1981.

[250] Sochen, N., Kimmel, R., and Malladi, R., *A General Framework for Low Level Vision*, IEEE Trans. Image Proc., 7,3, pp. 310–318, 1998.

[251] Sod, G.A., *Numerical Methods in Fluid Dynamics*, Cambridge University Press, 1985.

[252] Son, G., and Dhir,V.K., *Numerical Simulation of Film Boiling Near Critical Pressures with a Level Set Method*, J. Heat Transfer, 120, pp. 183–192, 1998.

[253] Souganidis, P.E., *Approximation Schemes for Viscosity Solutions of Hamilton–Jacobi Equations*, J. Diff. Eqns., 59, pp. 1–43, 1985.

[254] Strain, J., *Linear Stability of Planar Solidification Fronts*, Physica D, 30, pp. 297–320, 1988

[255] Strain, J., *A Boundary Integral Approach to Unstable Solidification*, J. Comp. Phys., 85, pp. 342–389, 1989.

[256] Strain, J., *Velocity Effects in Unstable Solidification*. SIAM Jour. Appl. Math., 50, pp. 1–15, 1990.

[257] Suen, C.Y., Nadal, C., Legault, R., Mai, T.A., and Lam, L., *Computer Recognition of Unconstrained Handwritten Numerals*, Proc. of the IEEE, 80, 7, pp. 1162–1180, 1992.

[258] Sullivan, J. M., Lynch, D. R., and O'Neill, K. O, *Finite Element Simulation of Planar Instabilities during Solidification of an Undercooled Melt*, Jour. Comp. Phys., 69, pp. 81–111, 1987.

[259] Sussman, M., and Fatemi, E., *An Efficient, Interface-Preserving Level Set Re-Distancing Algorithm and its Application to Interfacial Incompressible Fluid Flow*, preprint, 1995.

[260] Sussman, M., Fatemi, E., Smereka, P., and Osher, S., *An Improved Level Set Method for Incompressible Two-Phase Flows*, Computers and Fluids, 27, 5-6, pp. 663–80,1998.

[261] Sussman, M., Smereka, P., *Axisymmetric Free Boundary Problems*, preprint, 1998.

[262] Sussman, M., Smereka, P. and Osher, S.J., *A Level Set Method for Computing Solutions to Incompressible Two-Phase Flow*, J. Comp. Phys. 114, pp. 146–159, 1994.

[263] Taylor, J.E., Cahn, J.W., Handwerker, C.A., *Geometric models of crystal growth.*, Acta Metallurgica et Materialia, 40, 7, pp. 1443–1474, 1992.

[264] Teboul, S., Blanc-Féraud, L., Aubert, G., and Barlaud, M., *Variational Approach for Edge-Preserving Regularization Using Coupled PDE's*. IEEE Trans. Image Proc., 7,3, pp. 387–397, 1998.

[265] Terzopoulos, D., *Regularization of Inverse Visual Problems Involving Discontinuities*, IEEE Trans. on Patt. Anal. and Mach. Intell., 8, 2, pp. 413–424, 1986.

[266] Terzopoulos, D., Witkin, A., and Kass, M., *Constraints on Deformable Models: Recovering 3D Shape and Nonrigid Motion*, Artificial Intelligence, 36, pp. 91–123, 1988.

[267] Thompson, J., Warsi, Z.U.A., and Mastin, C.W., *Numerical Grid*

Generation, Foundations and Applications, North-Holland, Amsterdam, 1985.

[268] 3DGeo Corporation, *Computing and Imaging using Fast Marching Methods*, 3DGeo Corporation, Internal Report, June, 1998.

[269] Technology Modeling Associates, Three-Dimensional Photolithography Simulation with Depict 4.0, Technology Modeling Associates, Internal Documentation, January 1996.

[270] Terrain, Topography simulation for IC technology; Reference manual, Avant! Corporation, Fremont, CA, U.S.A., 1998.

[271] Toh, K.K.H., and Neureuther, A.R., *Three-Dimensional Simulation of Optical Lithography*, Proceedings SPIE, Optical/Laser Microlithography IV, 1463, pp. 356–367, 1991.

[272] Tomlin, C., *Hybrid Control in Air Traffic Managennent Systems*, Ph.D. Thesis, Electrical Engineering and Computer Sciences, University of California at Berkeley, 1998.

[273] Ulich, G., *Provably Convergent Methods for the Linear and Nonlinear Shape from Shading Problem*, J. Math. Imaging, 9, 1, pp. 69–82, 1998.

[274] Vemuri, B.C., and Malladi, R., *Constructing Intrinsic Parameters with Active Models for Invariant Surface Reconstruction*, IEEE Trans. on Patt. Anal. and Mach. Intell., 15, 7, pp. 668–681, 1993.

[275] Van de Vorst, G.A.L., *Modeling and Numerical Simulation of Viscous Sintering*, PhD. Thesis, Eindhoven University of Technology, Febodruk-Enschede, The Netherlands, 1994.

[276] van Trier, J., and Symes, W.W., *Upwind Finite-difference Calculations of Traveltimes*, Geophysics, 56, 6, pp. 812–821, 1991.

[277] Vidale, J., *Finite-Difference Calculation of Travel Times*, Bull. of Seism. Soc. of Amer., 78, 6, pp. 2062–2076, 1988.

[278] Vidale, J., *Finite-difference calculation of traveltimes in three dimensions*, Geophysics, 55, pp. 521–526, 1990.

[279] Weickert, J., ter Haar Romeny, B., and Viergever. M., *Efficient and Reliable Schemes for Non-linear Diffusion Filtering*, IEEE Trans. Image Proc., 7,3, pp. 398–410, 1998.

[280] Young, M.S., Lee, D., Lee., R., and Neureuther, A.R., *Extension of the Hopkins Theory of Partially Coherent Imaging to Include Thin-Film Interference Effects* SPIE Optical/Laser Microlithography VI, 1927, pp. 452–463, 1993.

[281] Zhang, H., Zheng, L.L., Prasad, V., and Hou, T., *A Curvilinear Level Set Formulation for Highly Deformable Free Surface Problems with Application to Solidification*, Numerical Heat Transfer, Part B, Vol. 34, pp. 1–20, 1997.

[282] Zhao, H-K., Chan, T., Merriman, B., and Osher, S., *A Variational Level Set Approach to Multiphase Motion*, Jour. Comp. Phys., 127, pp. 179–195, (1996).

[283] Zhao, H-K., Merriman, B., Osher, S., and Wang, L., *Capturing the Behaviour of Bubbles and Drops Using the Variational Level Set Approach*, J. Comp. Phys., 143, 2, pp. 495–518, 1998.

[284] Zhu, J., and Ronney, P.D., *Simulation of Front Propagation at Large Non-dimensional Flow Disturbance Intensities*, Comb. Sci. Tech., 100, pp. 183–201, 1995.

[285] Zhu, J., and Sethian, J.A., *Projection Methods Coupled to Level Set Interface Techniques*, J. Comp. Phys., 102, pp. 128–138, 1992.

[286] Zhu, J., and Sethian, J.A., *Tracking Two-Phase Flow Problems in Two and Three Dimensions*, in progress, 1998.

Index